# DIAMONDS

BY

SIR WILLIAM
CROOKES

HARPER &
BROTHERS
LONDON & NEW YORK

# ·DIAMONDS·

BY

SIR WILLIAM CROOKES

LL.D., D.Sc., F.R.S.

Foreign Sec. R.S., Hon. LL.D. (Birmingham), Hon. Sc.D. (Camb. and Dubl.), Hon. D.Sc. (Oxon. and Cape of Good Hope); Past Pres. Chem. Soc., Brit. Assoc., Inst. Elect. Eng., Soc. Psych. Res.; Hon. Mem. Roy. Phil. Soc. Glasgow, Roy. Soc. N.S.W., Pharm. Soc., Chem. Metall. and Mining Soc. of South Africa, Amer. Chem. Soc., Amer. Philos. Soc., Roy. Soc. Sci. Upsala, Deutsch. Chem. Gesell. Berlin, Psychol. Soc. Paris, "Antonio Alzate" Sci. Soc. Mexico. Sci. Soc. Bucharest, Reg. Accad. Zelanti, Aci Reale; Corresp. Inst. de France (Acad. Sci.), Corresp. Mem. Bataafsch Genoots. Rotterdam, Soc. d'Encouragement pour l'Indust. Paris, For. Mem. Accad. Lincei Rome

WITH 24 ILLUSTRATIONS

LONDON AND NEW YORK
HARPER & BROTHERS
45 ALBEMARLE STREET, W.
1909

TO MY WIFE

MY COMPANION AND FRIEND OF

FIFTY-FOUR YEARS.

TO HER JUDGMENT AND ADVICE I OWE MORE

THAN I CAN EVER REPAY

AND TO HER I DEDICATE THIS BOOK.

# PREFACE

THE following pages are based on personal observations during two visits to Kimberley, in 1896 and 1905, and on personal researches on the formation and artificial production of diamonds. In 1896 I spent nearly a month at Kimberley, when Mr. Gardner F. Williams, the General Manager of the De Beers Consolidated Mines, and the managers of neighbouring mines, did their utmost to aid in my zealous quest for reliable information. They gave me free access to all workings above and below ground, allowed me to examine at leisure their stock and to take extracts from their books. I had exceptional opportunities of studying the geology of the Diamond and of noting the strange cataclysmal facts connected with the birth, growth, and physics of the lustrous stones.

# PREFACE

In 1905 with my wife I returned to Kimberley. We were members of the British Association which held its meeting that year in South Africa. I was asked to give one of the Association lectures at Kimberley and it was natural for me to discourse "On Diamonds." During our stay we were the guests of Mr. Gardner Williams.

Returning to England after the visit of 1896, I gave two lectures on Diamonds at the Imperial Institute and one at the Royal Institution. These lectures, and the lecture delivered at Kimberley, in 1905—hitherto only privately distributed—form the basis of the present volume. On each visit I took abundant photographs, many of which I now reproduce. A few are copied from plans lent by Mr. Gardner Williams and one or two are from photographs purchased at Kimberley.

In obtaining statistical information of the Diamond industry, I owe much to the Annual Reports of the De Beers Company. I have also quoted freely from Reunart's valuable book on *Diamonds and Gold in South Africa*; and I render my acknowledgments to the authors of the following papers and memoirs.

*On a Visit to the Diamond Fields of South Africa, with Notices of Geological Phenomena by the Wayside.* By John Paterson, Esq., M.A.

*On the Mode of Occurrence of Diamonds in South Africa.* By E. J. Dunn.

*On the Origin and Present Position of the Diamonds of South Africa.* By G. G. Cooper, Esq., of Graaf Reinet.

*On the Character of the Diamantiferous Rock of South Africa.* By Prof. N. Storey Maskelyne, F.R.S., Keeper, and Dr. W. Flight, Assistant in the Mineral Department, British Museum.

*Further Notes on the Diamond Fields of South Africa.* By E. J. Dunn.

*Notes on the Diamond Fields of South Africa*, 1880. By E. J. Dunn.

*Analogies between the Diamond Deposits in*

*South Africa and those in Meteorites.* By M. Daubrée.

*Notes on the Diamond-bearing Rock of Kimberley, South Africa.* By Sir J. B. Stone, Prof. T. G. Bonney, and Miss Raisin.

*Notes on the Diamond Rock of South Africa.* By W. H. Hudleston.

*The Parent Rock of the Diamond in South Africa.* By the Reverend Professor T. G. Bonney.

The Presidential Address, by Grove Carl Gilbert, to the Geological Society of Washington, on *The Origin of Hypotheses. Illustrated by the Discussion of a Topographical Problem.* 1896.

*Le Four Electrique.* By Henri Moissan. 1897.

*The Diamond Mines of South Africa.* By Mr. Gardner F. Williams. (In this publication the story of the rise and development of the industry is exhaustively narrated.)

*British Association, South African Meeting,* 1896, *Kimberley Handbook.*

## PREFACE

*The Meteor Crater of Canyon Diablo, Arizona ; its History, Origin, and Associated Meteoric Irons.* By George P. Merrill. 1908.

In the present volume I have tried to give some idea of the underground wonders of the Kimberley mines. I have pictured the strenuous toil of the men who bring to the surface the buried treasures, and I have given some idea of the skill and ingenuity with which their labours are controlled. I have done my best to explain the fiery origin of the Diamond, and to describe the glowing, molten, subterranean furnaces where they first begin mysteriously to take shape. I have shown that a diamond is the outcome of a series of Titanic earth convulsions, and that these precious gems undergo cycles of fiery, strange, and potent vicissitudes before they can blaze on a ring or a tiara.

I am glad to have paid these two visits to South Africa. I always recall with interest the dusky smiling natives at work and at play. I am glad to have seen that Arabian Nights vision, the strong-room of the De Beers Company. Above all, I have vividly graven on my heart the friendly welcome, and the innumerable acts of kindness shown us by our able, energetic, and enterprising Colonial fellow-countrymen.

W. C.

# CONTENTS


| CHAPTER | | PAGE |
|---|---|---|
| I. | Preliminary | 1 |
| II. | Kimberley and its Diamond Mines | 14 |
| III. | Kimberley Mines at the Present Day | 34 |
| IV. | Collecting the Gems | 55 |
| V. | The Diamond Office | 73 |
| VI. | Noteworthy Diamonds | 76 |
| VII. | Boart, Carbonado, and Graphite | 81 |
| VIII. | Physical and Chemical Properties of the Diamond | 89 |
| IX. | Genesis of the Diamond | 115 |
| X. | The Natural Formation of the Diamond | 127 |
| XI. | Meteoric Diamonds | 134 |
| | Index | 145 |

# LIST OF PLATES

The Cullinan Diamond, from a photograph by the Author (see pp. 76-79) . . *Frontispiece*

| FIG. | | FACING PAGE |
|---|---|---|
| 1. | River Washings at Klipdam . . . | 10 |
| 2. | Plan of the Kimberley Diamond Mines . . | 10 |
| 3. | Kimberley Mine. The "Pipe" . . . | 18 |
| 4. | Section of Kimberley Mine . . . | 18 |
| 5. | Wesselton Diamond Mine. Open Workings . | 34 |
| 6. | De Beers Compound . . . . | 40 |
| 7. | De Beers Mine. Underground Workings . . | 40 |
| 8. | De Beers Washing and Concentrating Machinery | 48 |
| 9. | Sorting Concentrates for Diamonds. De Beers . | 54 |
| 10. | De Beers Diamond Office. 25,000 carats . . | 72 |
| 11. | De Beers Diamond Office. The Valuators' Table . . . . . | 72 |
| 12. | A group of large Diamond Crystals . . | 76 |
| 13. | Some Historic Diamonds . . . | 80 |
| 14. | Crystalline forms of native Diamonds . . | 86 |
| 15. | Triangular Markings on natural face of a Diamond Crystal . . . . | 88 |
| 16. | Triangular Markings artificially produced on a Diamond Crystal . . . . | 88 |

# LIST OF PLATES


| FIG. | | FACING PAGE |
|---|---|---|
| 17. | Diamond-cut Glass and Shavings . . . . | 98 |
| 18. | Diamonds in Röntgen Rays. A. Black Diamond in gold frame. B. Pink Delhi Diamond. C. Paste Imitation of B. . . . | 98 |
| 19. | Curve of Vapour Pressure of Carbon . *page* | 113 |
| 20. | Moissan's Electric Furnace . . . | 116 |
| 21. | Artificial Diamond made by the Author from molten iron . . . . . | 120 |
| 22. | Moissan's Artificial Diamonds . . . . | 120 |
| 23. | Diamonds from Canyon Diablo Meteorite . . | 138 |

# DIAMONDS

## CHAPTER I

### PRELIMINARY

FROM the earliest times the diamond has fascinated mankind. It has been a perennial puzzle—one of the "riddles of the painful earth." It is recorded in *Sprat's History of the Royal Society* (1667) that among the questions sent by order of the Society to Sir Philiberto Vernatti, Resident in Batavia, was one inquiring "Whether Diamonds grow again after three or four years in the same places where they have been digged out?" The answer sent back was, "Never, or at least as the memory of man can attain to."

In a lecture "On Diamonds," fifty years ago,* Professor Maskelyne said, "The dia-

* *Chemical News,* Vol. I, p. 208.

mond is a substance which transcends all others in certain properties to which it is indebted for its usefulness in the arts and its beauty as an ornament. Thus, on the one hand, it is the hardest substance found in nature or fashioned by art. Its reflecting power and refractive energy, on the other hand, exceed those of all other colourless bodies, while it yields to none in the perfection of its pellucidity." He was constrained to add, "The formation of the diamond is an unsolved problem."

Diamonds are found in widely separated parts of the globe. In the United States they have been found in Arkansas, where the work of testing the deposits is now going on steadily and quietly. The general geology and petrography of the area and the weathering of the peridotite are described in a paper read before the American Institute of Mining Engineers by Messrs. Kunz and Washington. In tests made with a diamond drill the peridotite was proved to depths of 200 feet. The green and yellow grounds underlying the layer of black, sticky soil are found to extend down 40 feet in places, and are estimated to average 20 feet in depth over the area. The outcrop of the peridotite is estimated to cover about 40 acres, and may be larger. Some 540 diamonds have been found, with an aggregate of 200 carats. The largest stone weighs about 6·5 carats, though the average size compares favourably with the general run of most of the South African mines. There is a large proportion of white stones, many of which are free from flaws and are very brilliant. The genuineness of the occurrence of diamonds in their matrix is again proved, one stone having been found imbedded in the green ground at a depth of 15 feet. This peridotite has the form of a volcanic pipe, and therefore its outcrop is limited to one place.

In California authentic finds of diamonds are recorded in Butte County, especially at

Cherokee, above Orville. These diamonds, however, have come from alluvial deposits and have been found generally in washing for gold. As yet no authenticated discovery of diamond in its original matrix in California is recorded.

In Brazil the diamond industry has been increasing of late years, and the old mines in the Diamantina country are being worked by American capital and by the American methods which have proved so successful at De Beers. It is estimated that the annual value of the diamonds exported from Brazil amounts to over £800,000, but it is impossible to arrive at accurate figures owing to the large quantities smuggled out of the country to avoid payment of the export tax.

British Guiana produces a small quantity of diamonds, mostly, however, of small size. Between January and September, 1907, 1564 carats were exported.

Indian diamonds chiefly come from the

states of Panna, Charkhari, and Ajaigarh. In 1905 India exported 3059 carats, valued at £5160.

## Cape Colony

It is a standing surprise to the watchful outsider how little attention is bestowed on some of our colonies. For instance, to the Cape Colony, comprising vast, varied, and productive regions, we have till recently manifested profound ignorance and consequent indifference. When the Cape Colony was first incorporated with the Empire, it was pronounced "a bauble, unworthy of thanks." Yet before the Suez Canal and the Waghorn overland route to India, the Cape, as commanding our road to India, Australia, and China, had a special importance. Even now it presents an alternative route which under conceivable circumstances may be of capital moment.

The high grounds above Cape Town are rich in medicinal health-giving waters. The districts where these springs occur are high-lying, free from malaria, and admirably adapted for the restoration of invalids. It needs only some distinguished power to set the fashion, some emperor, prince, or reigning beauty to take the baths and drink the waters, and the tide of tourists would carry prosperity to Aliwal North, Fraserburg, Cradock, and Fort Beaufort.

South Africa, as I shall endeavour to show in detail, is the most important source of diamonds on the earth, and ranks with Australia and California as one of the three great gold-yielding regions. But the wealth of South Africa is not only in its gold and diamonds. The province of Natal contains more coal than Britain ever owned before a single bucket had been raised, and the beds extend over the Orange River Colony, whilst valuable iron ores exist also in large quantities.

In the year 1896 I spent nearly a month at Kimberley. Mr. Gardner F. Williams,

General Manager of the De Beers Consolidated Mines, and the Managers of neighbouring mines, did their utmost to assist me in my inquiries and to ply me with valuable information. I had full access to all the workings, above and below ground, and was able to examine at leisure their stock and take extracts from their books.

Again, in the year 1905, I paid another visit to Kimberley as the guest of Mr. Gardner Williams on the occasion of the meeting of the British Association in South Africa.

## River Washings

Besides the matrix mines, where the stones are found in pipes supposed to be of volcanic origin, the alluvial deposits on the Vaal River are of considerable importance. The terraces and gravels along the Vaal River for about 200 miles have been worked for diamonds, the deposits sometimes extending several miles on each side of the river, and varying from a few inches to 40 or 50 feet in thickness. The diamonds are found almost everywhere through the gravel deposit.

Before describing the present mode of diamond extraction followed in the important mines, I will commence with these "River Washings," where, in their primitive simplicity, can be seen the modes of work and the simple machinery long since discarded in the large centres of the industry. The drift or so-called "river washings" present a very interesting phase of diamond industry. The work is carried on in the primitive fashion adopted in the early days of diamond discovery, every man working on his own little claim, assisted by a few natives, and employing primitive machinery (Fig. 1). The chief centre of the Vaal River washings is about 30 miles to the north-west of Kimberley, at a place called Klipdam No. 2. There was originally a Klipdam a few miles further, and here the miners congregated, but the exhaustion of their claims made them migrate to others not far off and reported to be richer. Here, accordingly, they re-erected their iron houses and called it Klipdam No. 2.

It is a mistake to speak of "river washings." The diamantiferous deposits are not special to the old or recent river bed, but appear to be alluvial deposits spread over a large tract of country by the agency of water, which at some period of time subsequent to the filling up of the volcanic pipes planed off projecting kopjes from the surface of the country and scattered the debris broadcast over the land to the north-west of Kimberley. The larger diamonds and other heavy minerals would naturally seek the lowest places, corresponding with the river bed, past and present. The fact that no diamonds are found in the alluvial deposits near Kimberley may perhaps be explained by supposing that the first rush was sufficiently strong to carry the debris past without deposition, and that deposition occurred when the stream slackened speed. At Klipdam No. 2 the diamantiferous earth is remarkably like river gravel, of a strong red colour—quite different from the Kimberley blue ground—and forms a layer from 1 to 8 feet thick, lying over a "hard pan" of amygdaloidal trap, the melaphyre of the Kimberley mines.

When I was at Klipdam the miners had congregated at a spot called "New Rush," where some good finds of diamonds had been reported. The gravel is dug and put into a machine resembling the gold miner's dolly, where it is rocked and stirred by rakes, with a current of water flowing over it. Here all the fine stuff is washed away and a rough kind of concentration effected. The residual gravel is put on a table and sorted for diamonds—an operation performed by the master. At one of the claims where work was proceeding vigorously I

FIG. I. RIVER WASHINGS AT KLIPDAM.

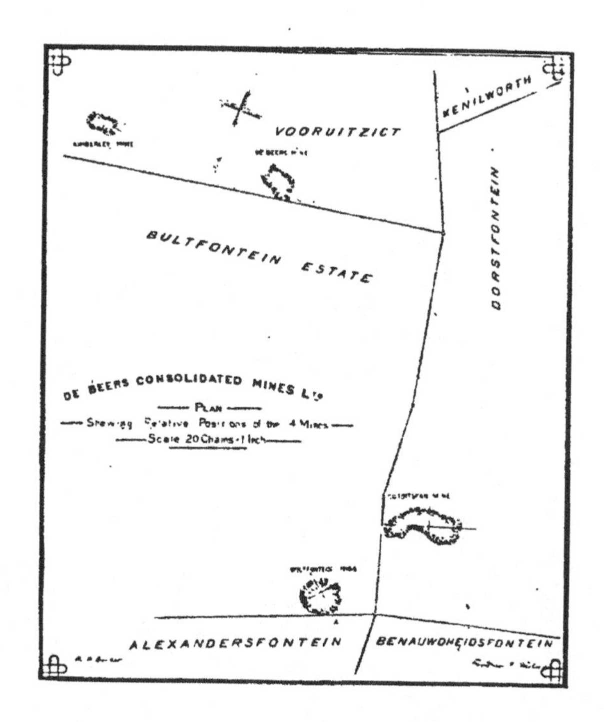


FIG. 2. PLAN OF THE KIMBERLEY DIAMOND MINES.

To face p. 10.

asked the proprietor to let me be present at the sorting out, as I should like to see river diamonds. He willingly consented, but no diamonds were to be found. On my expressing regret, he said he had not seen a diamond for a fortnight! I remarked that the prospect was rather a poor one, but he told me that a fortnight before he picked out one worth £300, "and that," he said, "will pay for several weeks' wages of my boys." This is the kind of speculative gambling that goes on at the river diggings. The miner may toil fruitlessly for months, and then come across a pocket of stones, where they have been swept by some eddy, by which he will net several thousands. Diamonds from the "river washings" are of all kinds, as if contributed by every mine in the neighbourhood. They are much rolled and etched, and contain a good proportion of first-class stones; they are of very good quality, as if only the better and larger stones had survived the ordeal of knocking about. Diamonds from the drift fetch about 40 per cent more than those from Kimberley; taking the yield of the Kimberley and De Beers mines as worth all round, large and small, 26s. 6d. a carat, those from the drift are worth 40s.

As a rule the better class of natives—the Zulus, Matabeles, Basutos, and Bechuanas—when well treated, are very honest and loyal to their masters. An amusing instance of the devotion of a Zulu came to my knowledge at Klipdam. He had been superintending a gang of natives on a small claim at the river washings. It yielded but few stones, and the owner—my informant—sold the claim, handing over the plant and small staff, our friend the Zulu remaining to look after the business till the new owner took possession. In the course of a few months the purchaser became dissatisfied with his bargain, not a single diamond having turned up since the transfer. One night the Zulu came to his old master in a mysterious manner, and laying a handful of diamonds on the table, said, "There, Baas, are your diamonds; I was not going to let the new man have any of them!"

# CHAPTER II

## KIMBERLEY AND ITS DIAMOND MINES

THE famous diamond mines in the neighbourhood are Kimberley, De Beers, Dutoitspan, Bultfontein, and Wesselton (Fig. 2). They are situated in latitude 28° 43′ South and longitude 24° 46′ East. Kimberley is practically in the centre of the present diamond-producing area. Besides these mines others of some importance of the Orange River Colony are known as Jaggersfontein and Koffyfontein, Lace, and Monastery, besides two new mines, the Roberts-Victor and the Voorspoed.

The areas of the mines are:

| | |
|---|---|
| Kimberley | 33 acres |
| De Beers | 22 acres |
| Dutoitspan | 45 acres |
| Bultfontein | 36 acres |

In 1907 the total number of carats raised from these mines was more than two million and a half, the sales of which realised £6,452,597.

The most important mine outside the Kimberley group is the new Premier Mine, about 20 miles West-North-West of Pretoria, where the famous Cullinan diamond was found.

Other diamond mines are the Frank Smith, Wesselton, the Kamfersdam, the Kimberley West, the Newlands, and the Leicester Mine.

The surface of the country round Kimberley is covered with a ferruginous red, adhesive, sandy soil, which makes horse traffic very heavy. Below the red soil is a basalt, much decomposed and highly ferruginous, from 20 to 90 feet thick, and lower still from 200 to 250 feet of black slaty shale containing carbon and iron pyrites. These are known as the Kimberley shales ; they are very combustible, and in a part of the De Beers Mine where they were accidentally fired they smouldered for over eighteen months. Then follows a bed of conglomerate about 10 feet thick, and below the conglomerate about 400 feet of a hard, compact rock of an olive colour, called "Melaphyre," or olivine diabase. Below the melaphyre is a hard quartzite about 400 feet thick. The strata are almost horizontal, dipping slightly to the north; in places they are distorted and broken through by protruding dykes of trap. There is no water nearer than the Vaal River, about 14 miles away, and formerly the miners were dependent on rain-water and a few springs and pools. Now, however, a constant and abundant supply of excellent water is served to the town, whilst good brick houses, with gardens and orchards, spring up on all sides. To mark the rate of progress, Kimberley has an excellent club and one of the best public libraries in South Africa. Parts of the town, affectionately called "the camp" by the older inhabitants, are not beyond the galvanised iron stage, and the general appearance is unlovely and depressing. Reunert reckons that over a million trees have been cut down to supply timber for the mines, and the whole country within a radius of 100 miles has been denuded of wood with the most injurious effects on the climate. The extreme dryness of the air, and the absence of trees to break the force of the wind and temper the heat of the sun, probably account for the dust storms so frequent in summer. The temperature in the day frequently rises to 100° in the shade, but in so dry a climate this is not unpleasant, and I felt less oppressed by this heat than I did in London the previous September. Moreover, in Kimberley, owing to the high altitude, the nights are always cool.

The approach to Kimberley is deadly dull. The country is almost treeless, and the bare veldt stretches its level length, relieved only by distant hills on the horizon.

### The Pipes or Craters

The five diamond mines or craters are all contained in a circle 3½ miles in diameter. They are irregularly shaped round or oval pipes, extending vertically downwards to an unknown depth, retaining about the same diameter throughout (Fig. 3). They are said to be volcanic necks, filled from below with a heterogeneous mixture of fragments of the surrounding rocks, and of older rocks such as granite, mingled and cemented with a bluish-coloured, hard clayey mass, in which famous blue clay the imbedded diamonds are hidden.

The craters or mines are situate in depressions, which have no outlets for the water which falls upon the neighbouring hills. The watersheds of these hills drain into ponds, called pans or vleis. The water, which accumulates in these ponds during

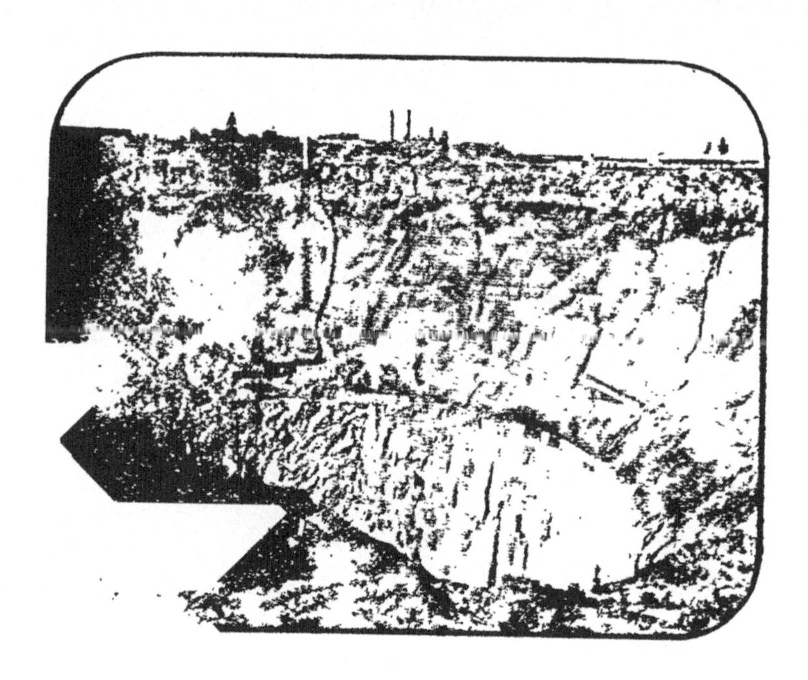

FIG. 3. KIMBERLEY MINE. THE "PIPE."

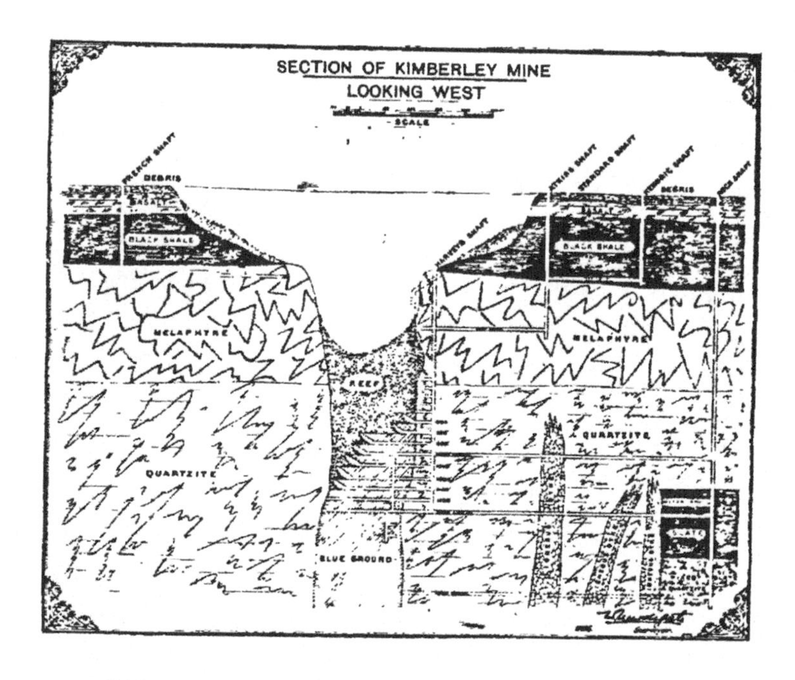


FIG. 4. SECTION OF KIMBERLEY MINE.

To face p. 18.

the rainy season, evaporates during the dry months, only one of them holding water throughout the dry season. The rocks which surround the craters are capped by red soil or calcareous tufa, and in places by both, the red soil covering the tufa.

The diamantiferous breccia filling the mines, usually called "blue ground," is a collection of fragments of shale, various eruptive rocks, boulders, and crystals of many kinds of minerals. Indeed, a more heterogeneous mixture can hardly be found anywhere else on this globe. The ground mass is of a bluish green, soapy to the touch and friable, especially after exposure to the weather. Professor Maskelyne considers it to be a hydrated bronzite with a little serpentine.

The Kimberley mine is filled for the first 70 or 80 feet with what is called "yellow ground," and below that with "blue ground" (Fig. 4). This superposed yellow on blue is common to all the mines. The blue is the unaltered ground, and owes its colour chiefly to the presence of lower oxides of iron. When atmospheric influences have access to the iron it is peroxidised and the ground assumes a yellow colour. The thickness of yellow earth in the mines is therefore a measure of the depth of penetration of air and moisture. The colour does not affect the yield of diamonds.

Besides diamonds, there have been detected more than eighty species of minerals in the blue ground, the more common being magnetite, ilmenite, garnet, bright green ferriferous enstatite (bronzite), a hornblendic mineral closely resembling smaragdite, calc-spar, vermiculite, diallage, jeffreysite, mica, kyanite, augite, peridot, eclogite, iron pyrites, wollastonite, vaalite, zircon, chrome iron, rutile, corundum, apatite, olivine, sahlite, chromite, pseudobrookite, perofskite, biotite, and quartz. The blue ground does not show any signs of passing through great heat, as the fragments in the breccia are not fused at the edges. The eruptive force was probably steam or water-gas, acting under great pressure, but at no high temperature. According to Mr. Dunn, in the Kimberley Mine, at a depth of 120 feet, several small fresh-water shells were discovered in what appeared to be undisturbed material.

A selection of thin sections of some of these rocks and minerals, mounted as microscopic objects and viewed by polarised light, are not only of interest to the geologist, but are objects of great beauty.

The appearance of shale and fragments of other rocks testify that the *mélange* has suffered no great heat in its present condition, and that it has been erupted from great depths by the agency of water vapour or some similar gas.

The rock outside the pipes and encasing them is called "reef." Inside some of the mines occur large masses of "floating reef," covering an area of several thousand square feet. In the De Beers Mine is what is called "the snake," a dyke of igneous rock taking a serpentine course across the mine, and standing like a vein nearly vertical, varying in thickness from 2 to 7 feet. The main body of the blue ground is entirely analogous to the snake rock, naturally more decomposed, but in essential points the microscopic appearance of the blue ground and of the "snake" is in an extraordinary degree alike. Mr. Gardner Williams supposes that the "snake" is a younger eruptive formation coming from the same volcanic source as the blue ground. No diamonds have been found either in the "snake" or the floating reef. The ground, however, is generally richer in diamonds in the neighbourhood of the floating reef.

Before the discovery of the mines there was nothing in the superficial appearance of the ground to indicate the treasures below. Since the volcanic ducts were filled with the diamantiferous ground, denudation has planed the surface and the upper parts of the craters, and other ordinary signs of volcanic activity being smoothed away, the superficial and ubiquitous red sand covered the whole surface. The Kimberley Mine seems to have presented a slight elevation above the surrounding flat country, while the sites of other mines were level or even slightly depressed. The Wesselton Mine, within a mile of Dutoitspan, has only been discovered a few years. It showed a slight depression on the surface, which had been used as a shoot for dry rubbish. There are other diamantiferous pipes in the neighbourhood, but they are small and do not contain stones in payable quantities. More recently another diamantiferous pipe has been discovered about 40 miles off, near Klipdam, and is now worked as the Leicester Mine. Other hoards of diamonds may also be near; where there are no surface signs, and the pipe itself is hidden under 10 or 20 feet of recent deposits, it is impossible to prospect the entire country. Accident has hitherto been the chief factor in the discovery of diamond mines.

How the great pipes were originally formed is hard to say. They were certainly not burst through in the ordinary manner of volcanic eruption, since the surrounding and enclosing walls show no signs of igneous action, and are not shattered or broken up even when touching the "blue ground." It is pretty certain these pipes were filled from below after they were pierced and the diamonds were formed at some previous time and mixed with a mud volcano, together with all kinds of debris eroded from the rocks through which it erupted. The direction of flow is seen in the upturned edges of some of the strata of shale in the walls, although I was unable to see any upturning in most parts of the walls of the De Beers Mine at great depths.

# KIMBERLEY IN OLD DAYS

## THE KIMBERLEY MINE IN OLD DAYS

According to Mr. Paterson, who examined the diamond fields of Kimberley soon after their discovery, "Wherever the diamond is obtained perfect in form and smooth in finest smoothness of surface, without depression, hump, or twist of any kind, such diamonds were ever found in their own little moulds of finest limey stuff,* and as if such mould of lime had been a necessity to their perfect formation. And further, where the splinters of diamonds, or boarty stuff, were chiefly met by the diggers, there was much less presence of limey matter in the claim at the section of it where such broken or fragmentary diamonds were found; and that chiefly from among what the diggers termed 'clay-ballast,' or 'burnt brick,' were un-

* Mr. Paterson called "limey stuff" what is now termed "blue ground." It was also formerly called "marl stuff," "blue stuff," and "blue clay."

earthed the bits or undeveloped crystals so plentiful at New Rush." *

In the first days of diamond mining there was no idea that diamantiferous earth extended to any particular depth, and miners were allowed to dig holes at haphazard and prospect where they liked. When the Kimberley Mine was discovered a new arrangement was made, and in July, 1871, it was cut up into about 500 claims, each 31 feet square, with spaces reserved for about fifteen roadways across the mine. No person at first could hold more than two claims—a rule afterwards modified.

The following quotation from a description of a visit to Kimberley in 1872, by Mr. Paterson, taken from a paper read by him to the Geologists' Association, gives a graphic picture of the early days of the Kimberley Mine:

"The New Rush diggings (as the Kimber-

* The original name for the Kimberley Mine. It was also sometimes known as "Colesberg Kopje."

ley Mine was at first called) are all going forward in an oval space enclosed around by the trap dyke, and of which the larger diameter is about 1000 feet, while the shorter is not more than 700 feet in length. Here all the claims of 31 feet square each are marked out with roadways of about 12 feet in width, occurring every 60 feet. Upon these roadways, by the side of a short pole fixed into the roadway, sits the owner of the claim with watchful eye upon the Kafir diggers below, who fill and hoist, by means of a pulley fixed to the pole above, bucketful after bucketful of the picked marl stuff in which the diamonds are found.

"Many of the claims are already sunk to a depth of 100 feet, and still the diamonds continue to be found as plentifully as ever. From the roadway above the marl is carted away to the sorting-tables, outside the range of the diggings, among mounds of marl stuff which seem like little hills. Here, amidst such whirls of dust as are nowhere else seen, the marl stuff is pounded, sifted from the finest powder of lime and clay, and from the residue put on the sorting-tables, the diggers, with a piece of zinc 9 inches long by 4 inches in breadth, search out in the successive layers taken from the heap the precious gems. I need not tell you that the search is by no means very perfect, or that perhaps as many diamonds escape the digger's eye as are discovered and taken out by him, but you will perhaps confess with me that their aptness in picking out the diamonds is by no means to be despised, when I tell you that in one six months from the date of opening New Rush diggings, little short of a million sterling in diamonds has been extracted from them. At close of day the diggers take daily stock of their finds, and between five and six o'clock in the afternoon are to be seen hundreds and hundreds moving through the main street of New

Rush on visits to the tents of the buyers, seated behind their little green baize tables, with scales all ready, and bags of gold and silver and piles of banknotes, to buy the little gems."

It may help to realise the enormous value of the Kimberley Mine if I say that two claims, measuring together 62 by 31 feet and worked to a depth of 150 feet, yielded 28,000 carats of diamonds.

The roadways across the mine soon, however, became unsafe. Claims were sunk 100 or 200 feet each side of a roadway, and the temptation to undermine roadways was not always resisted. Falls of road frequently took place, followed by complete collapse, burying mine and claims in ruin. At that time there were probably 12,000 or 15,000 men at work in the mine, and then came the difficulty how to continue working the host of separate claims without interference with each other. A system of rope haulage was adopted.

## DIAMONDS

The following description of the work at the Kimberley Mine at this stage of its history is given by Mr. Reunert: *

"A succession of tall, massive timber stagings was erected round the margin of the mine. Each staging carried two or three platforms one above the other, every platform serving as an independent level from which to communicate with the claims below. Stationary ropes were then stretched from the different levels of the stagings to the claims, the ropes being anchored to the ground at both ends: the upper platforms communicated with the claims in the centre of the mine, the lower platforms with those nearer the margin. The hauling ropes were attached to windlasses worked by Kafirs on the several platforms, on which grooved guide wheels for the ropes were also fixed, the buckets being swung from the stationary ropes

* *Diamonds and Gold in South Africa.* By T. Reunert. Johannesburg, 1893.

by little overhead runners and crooks. Arrived at the level of the platform the bucket was tipped into a narrow shoot, down which the ground ran into a bag held ready to receive it, in which it was conveyed away to be sorted. The din and rattle of these thousands of wheels and the twang of the buckets along the ropes were something deafening, while the mine itself seemed almost darkened by the thick cobweb of ropes, so numerous as to appear almost touching. This mode of haulage continued in vogue during the whole of 1873, and if the appearance of the mine was less picturesque than when the roadways existed, it was, if anything, more unique. By moonlight, particularly, it was a weird and beautiful sight."

The mine was now threatened in two other quarters. The removal of the blue ground took away the support from the walls of the pipe, and frequent falls of reef occurred, not only covering up valuable claims with rubbish, but endangering the lives of workers below. Moreover, as the workings deepened, water made its appearance, necessitating pumping. In 1878 one quarter of the claims were covered by reef, and in 1879 over £300,000 were spent on removing reef and water. In 1881 over £200,000 were thus spent, and in 1882 more than half a million sterling was needed to defray the cost of reef removal. So matters went on until four million cubic yards of reef had been removed, at a cost of two millions sterling, and still little good was done, for out of 400 claims in the mine only about fifty could be regularly worked. Ultimately, in November, 1883, the biggest fall of reef on record took place, estimated at 250,000 cubic yards, surging half across the mine, where the bulk of it lies to this day. It became evident that open workings could not be carried on at such depths, and after many experiments the present system of underground working was devised.

During this time of perplexity, individual miners who could easily have worked one or two claims near the surface could not continue work in the face of harassing difficulties and heavy expenses. Thus the claims gradually changed hands until the mine became the property first of a comparatively small number of capitalists, then of a smaller number of limited liability companies, until finally the whole of the mines have practically become the property of the "De Beers Consolidated Mines, Limited."

# CHAPTER III

## KIMBERLEY MINES AT THE PRESENT DAY

THE De Beers Consolidated Mines, Limited, was founded in 1888, mainly through the genius of the late Cecil John Rhodes, for the purpose of acquiring all-important diamond-mining interests in the Kimberley area and thereby controlling the output. The two richest mines, Kimberley and De Beers, have been actively worked ever since, and have been the main contributors to an output which now realises over five millions sterling annually. Dutoitspan Mine was completely closed down, and practically the whole of Bultfontein was kept idle for many years; but with a view to the requirements of the future and the marked increase in the demand for diamonds, notwithstanding the

FIG. 5. WESSELTON DIAMOND MINE. OPEN WORKINGS.

To face p. 34.

steady rise in prices that has taken place, both these mines have now been equipped for underground working on a grand scale. The youngest of the De Beers group of mines is the Wesselton, which was discovered in 1890 by the late Mr. H. A. Ward, and soon afterwards purchased by Mr. Rhodes on behalf of the Company. The mine is now being worked opencast on a magnificent scale and has largely exceeded original expectations (Fig. 5). The success of the consolidation is proved by the fact that since it was brought about £22,000,000 have been paid in dividends to the shareholders, and it is roughly estimated that 40,000,000 carats of diamonds have been produced of a total value of eighty millions.

At the four mines about 8000 persons are daily employed, namely, 1500 whites and 6500 blacks. The wages are, whites, £5 or £6 a week; blacks, underground, 4s. to 5s. a day, and aboveground, 21s. a week.

## The Compound System

With gems like diamonds, where so large an intrinsic value is concentrated into so small a bulk, it is not surprising that robbery has to be guarded against in the most elaborate manner. The Illicit Diamond Buying (I.D.B.) laws are very stringent, and the searching, rendered easy by the "compounding" of the natives—which I shall describe presently—is of the most drastic character (Fig. 6). It is, in fact, very difficult for a native employee to steal diamonds; even were he to succeed, it would be almost impossible to dispose of them, as a potential buyer would prefer to secure the safe reward for detecting a theft rather than run the serious risk of doing convict work on the Cape Town Breakwater for a couple of years. I heard of a native who, secreting a diamond worth several hundreds of pounds, after trying unsuccessfully to sell it, handed it back to the manager of his compound, glad to get the sixpence a carat to which he was entitled. Before the passing of the "Diamond Trade Act" the value of diamonds stolen reached nearly one million sterling per annum.

A "compound" is a large enclosure about 20 acres in extent, surrounded by rows of one-story buildings of corrugated iron. These are divided into rooms holding each about twenty natives. A high iron fence is erected around the compound, 10 feet from the buildings. Within the enclosure is a store where the necessaries of life are supplied to the natives at a reduced price, wood and water being provided free of charge. In the middle is a large swimming-bath, with fresh water running through it. The rest of the space is devoted to recreation, games, dances, concerts, and any other amusement the native mind can desire. I have to thank the superintendents of the respective compounds, who spoke all the native dialects, for their kindness in showing us round, and suggesting dances and concerts, got up at ten minutes' notice, for the benefit of my camera. The dancing was more of the character of attitudinising and marching to a monotonous tum-tum, the "orchestra" consisting of various-sized drums and what they call a piano—an octave or so of tuned slabs of wood held in order on stretched strings and struck with a wooden hammer. The native music as a rule is only marking time, but I have heard musical melodies accompanying some of their songs. In case of accident or illness there is a well-appointed hospital where the sick are tended. Medical supervision, nurses, and food are supplied free by the Company.

In the compound are to be seen representatives of nearly all the picked types of African tribes. Each tribe keeps to itself, and to go round the buildings skirting the compound is an admirable object-lesson in ethnology. At one point is a group of

Zulus; next we come to Fingoes; then Basutos; beyond come Matabele, Bechuanas, Pondos, Shangains, Swazis, and other less-known tribes, either grouped or wandering around making friendly calls.

The clothing in the compound is diverse and original. Some of the men are evident dandies, whilst others think that in so hot a climate a bright-coloured handkerchief or "a pair of spectacles and a smile" is as great a compliance with the conventions of civilisation as can be expected.

The natives are not interfered with in their various amusements, always provided they do not make themselves objectionable to their neighbours. They soon learn that tribal animosities are to be left outside the compound. One Sunday afternoon my wife and I walked unattended about the compound, almost the only whites present among 1700 natives. The manners of the fold were so friendly, and their smiles so cordial, that the idea of fear vanished. At one part a Kafir was making a pair of trousers with a bright nickel-plated sewing-machine, in which he had invested his savings; next to him a "boy" was reading from the Testament in his own language to an attentive audience; in a corner a party were engaged in cooking a savoury mess in an iron pot; further on the orchestra was tuning up and Zulus were putting the finishing touches to their toilet of feathers and beads. One group was intently watching a mysterious game. It is played by two sides, with stones and grooves and hollows in the ground, and appears of most absorbing interest. It seems to be universal throughout Africa; it is met with among the ruins of Zimbabwe, and signs of it are recorded on old Egyptian monuments. I wanted to learn it, and an intelligent Zulu player offered to teach it to me in a few minutes. Captain Dallas, however, with a more accurate opinion of my intelligence than my friend the Zulu, assured me it

FIG. 6. DE BEERS COMPOUND.

FIG. 7. DE BEERS MINE. UNDERGROUND WORKINGS.

To face p. 40.

would take months before I could begin to know anything about it. He had tried for years and could make nothing of it.

They get good wages, varying according to occupation. The work is appreciated, and there are always more applicants than can be accepted. On entering, the restrictions to which they must submit are fully explained, and they are required to sign for three months at least, during which time they must not leave the compound or mine. A covered way and tunnel lead the workers underground to the down shaft, while those working on the depositing floors go and come under guard. It is seldom that a man does not return once he has lived the life in the compound; some come again and again for years, only leaving occasionally to spend accumulated savings. The most careful men save money, and carry it at intervals to the superintendent to keep for them. Occasionally they ask to look at their savings, which may amount to £30 or

£40, accumulated by driblets. They are ignorant of savings banks or interest, and are content if they see their own money in the original rags and papers. The Kafir, on demand, must behold his coins just as he handed them in, wrappings and all. Sometimes the superintendent will have as much as £1000 of savings in his care.

On leaving, the men generally draw all their savings, and it is not uncommon for a grateful Kafir to press £2 or £3 on Captain Dallas in recognition of his trouble. They are astonished when their offerings are declined; still more so when it is explained that if they would put their savings in a bank they would have a few extra pounds given to them for the privilege of taking care of it.

A shrewd young Pondo, who had been coming year after year, applied for some of his savings, and gave as a reason that he wanted to buy a wife. "But you said the same thing last year," replied Captain

Dallas ; "I hope nothing has happened." "No," said the man ; "one wife, she quarrel with me ; two wives, they quarrel with each other ; me peace !"

## Underground Workings

In the face of constant developments I can only describe the system in use at the time of my own visits in 1896 and 1905. Shafts are sunk in the solid rock at a sufficient distance from the pipe to be safe against reef movements in the open mine. In 1903 the rock shafts in the De Beers and Kimberley Mines reached depths of 2076 and 2599 feet respectively. Tunnels are driven from these shafts at different levels, about 120 feet apart, to cross the mine from west to east. These tunnels are connected by two other tunnels running north and south, one near the west side of the mine and one midway between it and the east margin of the mine. From the east and west tunnels offsets are driven to the surrounding rock. When near the rock the offsets widen into galleries, these in turn being stoped on the sides until they meet, and upwards until they break through the blue ground. The fallen reef with which the upper part of the mine is filled sinks and partially fills the open space. The workmen then stand on the fallen reef and drill the blue ground overhead, and as the roof is blasted back the debris follows. When stoping between two tunnels the blue is stoped up to the debris about midway between the two tunnels. The upper levels are worked back in advance of the lower levels, and the works assume the shape of irregular terraces. The main levels are from 90 to 120 feet apart, with intermediate levels every 30 feet. Hoisting is done from only one level at a time through the same shaft. By this ingenious method every portion of blue ground is excavated and raised to the surface, the rubbish on the top gradually sinking and taking its place.

## UNDERGROUND WORKINGS

The scene below ground in the labyrinth of galleries is bewildering in its complexity, and very unlike the popular notion of a diamond mine (Fig. 7). All below is dirt, mud, grime; half-naked men, dark as mahogany, lithe as athletes, dripping with perspiration, are seen in every direction, hammering, picking, shovelling, wheeling the trucks to and fro, keeping up a weird chant which rises in force and rhythm when a greater task calls for excessive muscular strain. The whole scene is more suggestive of a coal mine than a diamond mine, and all this mighty organisation, this strenuous expenditure of energy, this costly machinery, this ceaseless toil of skilled and black labour, goes on day and night, just to win a few stones wherewith to deck my lady's finger! All to gratify the vanity of woman! "And," interposed a lady who heard this remark, "the depravity of man!"

## The Depositing Floors

Owing to the refractory character of blue ground fresh from the mines, it has to be exposed to atmospheric influences before it will pulverise under the action of water and mechanical treatment.

From the surface-boxes, into which the blue ground is tipped when it reaches the top of the main shaft, it is transferred to side-tipping trucks and sent to the depositing floors by means of endless wire-rope haulage. The speed of the haulage varies from 2½ to 4 miles per hour. The trucks are counted automatically as they are sent to the floor by a reciprocating engine-counter placed on a frame near the tram-line.

The depositing floors are prepared by removing the bush and grass from a fairly level piece of ground; this ground is then rolled smooth and hard. The floors extend over many square miles of country and are surrounded by 7-foot barbed wire fences, vigilantly guarded day and night. The De Beers floors, on Kenilworth, are laid off in rectangular sections 600 yards long and 200 yards wide, each section holding about 50,000 loads. The ground from the Kimberley Mine is the softest and only needs a few months' exposure on the floors; the ground from De Beers is much harder and requires at least six months' exposure, while some ground is so hard that it will not disintegrate by exposure to the weather under one or two years. The De Beers Mine contains a much larger quantity of this hard blue ground than the other mines, and in order to save the loss of time consequent on keeping an enormous stock of blue constantly on the floors, it has recently been decided to pass the harder and more refractory stuff direct from the mine through crushing mills.

For a time the blue ground remains on the floors without undergoing much alteration. But soon the heat of the sun and moisture produce a wonderful effect. Large pieces, hard as ordinary sandstone when taken from the mine, commence to crumble. At this stage the winning of the diamonds assumes more the nature of farming than mining. The ground is frequently harrowed and occasionally watered, to assist pulverisation by exposing the larger pieces to atmospheric influences. The length of time necessary for the ground to weather before it becomes sufficiently pulverised for washing depends on the season of the year and the amount of rain. The longer the ground remains exposed the better it is for washing.

It is curious to note that there is a marked difference in the rapidity of disintegration of the blue ground in each of the four mines. The longer the exposure, the more complete the pulverisation and the better for washing. Under normal conditions soft blue ground becomes sufficiently

FIG. 8. DE BEERS WASHING AND CONCENTRATING MACHINERY.

To face p. 48.

pulverised in from four to six months, but it is better to expose it for a longer period, even for a whole year.

### Washing and Concentrating Machinery

After the blue ground has been weathered for a sufficient time, it is again loaded into trucks and hauled to the crushing machinery (Fig. 8). The first or "comet" crushers reduce the ground so that it will pass into hoppers and thence into revolving cylinders covered with perforated steel plates, having holes 1¼ inches in diameter which separate the finely crushed from the coarse pieces.

Pieces larger than 1¼ inches pass out of the end of the cylinders and fall upon a conveyor belt, which takes them to the end of the machine—these pieces are mostly waste rock which is found in the blue ground.

The fine ground which passes through the holes in the cylinder, together with a plentiful current of water, flows into the washing pans. These pans are of iron,

14 feet in diameter, furnished with ten arms each having six or seven teeth. The teeth are so set as to form a spiral, so that when the arms revolve the teeth carry the heavy deposit to the outer rim of the pan, while the lighter material passes towards the centre and is carried from the pan by the flow of water. The heavy deposit contains the diamonds. It remains on the bottom of the pan and near its outer rim. This deposit is drawn off every twelve hours by means of a broad slot in the bottom of the pan. The average quantity of blue ground passed through each pan is from 400 to 450 loads in ten hours. The deposit left in each pan after putting the above number of loads through amounts to three or four loads, which go to the pulsator for further concentration.

About 14 per cent of all the ground sent to the depositing floors is too hard to weather, so of late years crushing and concentrating plant has been erected to deal effectually with the hard lumps, thus saving the great lock-up of capital consequent on letting them lie on the floor a year or two.

The hard lumps being hauled to the upper part of the machine, are tipped into bins, whence they pass to crushing rollers which so reduce them that they will pass through a ring two inches in diameter. The coarse powder is screened through revolving cylinders having ½-inch and 1¼-inch perforations. The stuff passing through the finer holes goes to the finishing mill, while the coarser stuff goes to smaller crushers. Before the coarse lumps are re-crushed they pass over revolving picking tables, where any specially large diamonds are rescued, thus preventing the risk of breakage. From the picking tables the ground is scraped automatically into two sets of rolls, and the pulverised product screened again and graded into three sizes. The finest size, passing a ½-inch screen, goes to the washing pans, and the two coarser sizes to jigs. Large diamonds which have been separated from their envelope of blue are retained in the jig. The ground still holding the smaller diamonds passes out of the end of the jig and then through a series of rolls, screens, and jigs until the diamantiferous gravel is drawn from the bottom jigs into locked trucks running on tramways to the pulsator for further concentration and sorting.

The pulsator is an ingeniously designed but somewhat complicated machine for dealing with the diamantiferous gravel already reduced one hundred times from the blue ground, the pulsator still further concentrating it till the gravel is rich enough to enable the stones to be picked out by hand. The value of the diamonds in a load of original blue ground being about 30s., the gravel sent to the pulsator from the pans, reduced a hundredfold, is worth £150 a load. Stuff of this value must not be exposed to risk of peculation.

## THE PULSATOR

The locked trucks are hoisted by a cage to a platform, where they are unlocked and their contents fed into a shoot leading to a cylinder covered with steel sieving with holes from $\frac{1}{16}$ to $\frac{5}{8}$ of an inch in diameter. The five sizes which pass through the cylinder flow upon a combination of jigs, termed at the mines the pulsators. The bottoms of the jigs are covered with screens, or sieving, the meshes of which are a little larger than the holes in the revolving cylinder immediately at the back of them.

Over each screen is spread a layer of bullets to prevent the rich deposit from passing too rapidly through the screens. The jigs themselves are stationary, but from below an intermittent stream of water passes in rapid pulsations with an up and down movement. This pulsation keeps the diamantiferous gravel constantly moving —"alive" is the expressive word used—and tends to sort out the constituents roughly according to their specific gravity, the heavier particles working to the bottom and the lighter material washing off by the flow of water and passing into trucks, whence it is carried to the tailings heap. The heavier portions, by the up and down wash of the water, gradually work their way under the bullets and pass through the screens into pointed boxes, whence the heavy concentrates are drawn off upon endless belts. These convey their precious load to small elevators by means of which the concentrates are lifted into hoppers from which they are fed upon shaking tables.

FIG. 9. SORTING CONCENTRATES FOR DIAMONDS. DE BEERS.

To face p. 54.

# CHAPTER IV

## COLLECTING THE GEMS

THE sorting room in the pulsator house is long, narrow, and well lighted (Fig. 9). Here the rich gravel is brought in wet, a sieveful at a time, and is dumped in a heap on tables covered with iron plates. The tables at one end take the coarsest lumps, next comes the gravel which passed the ⅜-inch holes, then the next in order, and so on. The first sorting is done by thoroughly trustworthy white men; for here the danger of robbery is greatest. Sweeping the heap of gravel to the right, the sorter scrapes a little of it to the centre of the table by means of a flat piece of sheet zinc. With this tool he rapidly passes in review the grains, seizes the diamonds and puts them into a little tin box in front of him. The stuff is then swept off to the left and another lot taken, and so on till the sieveful of gravel is exhausted, when another is brought in. The stuff the sorter has passed to his left as temporarily inspected is taken next to another part of the room, where it is again scrutinised by native convicts again and again, and whilst diamonds can be found in quantity sufficient to repay the cost of convict labour, it is passed under examination.

The diamond has a peculiar lustre, and on the sorter's table it is impossible to mistake it for any other stone that may be present. It looks somewhat like clear pieces of gum arabic, with a sort of intrinsic lustre which makes a conspicuous shine among the other stones.

### Automatic Diamond Collector

A series of experiments was initiated by Mr. Gardner Williams with the object of separating the diamonds from the heavy, valueless concentrates with which they are associated. An ordinary shaking or percussion table was constructed, and every known means of separation was tried without success. One of the employees of De Beers, Mr. Fred Kirsten, was in charge of the experimenting, under the supervision of the late Mr. George Labram, the manager of the large crushing plant, and afterwards mechanical engineer to the Company. Notwithstanding the fact that the specific gravity of the diamond (3·52) was less than that of several of the minerals associated with it, so that its separation would seem a simple matter, it was found in practice to be impossible owing to the slippery nature of the diamond. The heavy concentrates carried diamonds, and diamonds flowed away from the percussion table with the tailings. When it seemed that every resource to do away with hand-sorting had been exhausted, Kirsten asked to be allowed to try to catch the diamonds by placing a coat of thick grease on the surface of the percussion table with which the other experiments had been made. Kirsten had noticed that oily substances, such as axle grease and white or red lead, adhered to diamonds when they chanced to come into contact, and, he argued to himself, if these substances adhered to diamonds and not to the other minerals in the concentrates, why should not diamonds adhere to grease on the table and the other minerals flow away? In this way the remarkable discovery was made that diamonds alone of all minerals contained in the blue ground will adhere to grease, and that all others will flow away as tailings over the end of the percussion table with the water. After this was determined by thorough experiments, more suitable shaking tables were constructed at the Company's workshops. These were from time to time improved upon, until now all the sorting (except for the very coarse size) is done by these machines, whose power of distinction is far superior to the keenest eye of the native.

Only about $\frac{1}{3}$ of 1 per cent of diamonds is lost by the first table, and these are recovered almost to a stone when the concentrates are passed over the second table. The discrimination of this sorter is truly marvellous. Native workers, although experienced in the handling of diamonds, often pick out small crystals of zircon, or Dutch boart, by mistake, but the senseless machine is practically unerring.

The grease containing the diamonds, together with a small percentage of very heavy minerals, such as iron pyrites and barytes, is scraped from the tables, placed in buckets made of steel plates with fine perforations, and boiled or steamed. The grease passes away to tanks of water, where it is cooled and is again fit for use. The diamonds, together with small bits of iron pyrites, brass nails from the miners' boots, pieces of copper from the detonator used in blasting, which remain on the tables owing to their high specific gravity, and a very small admixture of worthless deposit which has become mechanically mixed with the grease, are then boiled in a solution containing caustic soda, where they are freed from all grease. The quantity of deposit from the size of $\frac{5}{8}$ of an inch downwards, which now reaches the sorting table, does not exceed 1 cubic foot for every 12,000 loads (192,000 cubic feet) of blue ground washed. As already stated, $\frac{5}{12}$ of 1 per cent of the whole mass of blue formerly passed to the sorting tables; or, from 12,000 loads, which is about the daily average of the quantity washed at De Beers and Kimberley Mines, 800 cubic feet had to be assorted by hand.

### The Yield of Diamonds

Sometimes as many as 8000 carats of diamonds come from the pulsator in one day, representing about £20,000 in value.

When the bare statement is made that

nearly 5,000,000 truck-loads, or more than 4,000,000 tons of blue ground, have been washed in a year, the mind only faintly conceives the prodigious size of the mass that is annually drawn from the old craters and laboriously washed and sorted for the sake of a few bucketfuls of diamonds. It would form a cube of more than 430 feet, or a block larger than any cathedral in the world, and overtopping the spire of St. Paul's, while a box with sides measuring 2 feet 9 inches would hold the gems. From two to three million carats of diamonds are turned out of the De Beers mines in a year, and as 5,000,000 carats go to the ton, this represents half a ton of diamonds. To the end of 1892 10 tons of diamonds had come from this mine, valued at £60,000,000 sterling. This mass of blazing diamonds could be accommodated in a box 5 feet square and 6 feet high.

The diamond is a luxury, and there is only a limited demand for it throughout the world.

From four to four and a half millions sterling is as much as is spent annually in diamonds; if the production is not regulated by the demand, there will be over-production, and the trade will suffer. By regulating the output the directors have succeeded in maintaining prices since the consolidation in 1888.

The blue ground varies in its yield of diamonds in different mines, but is pretty constant in the same mine. In 1890 the yield per load of blue ground was:

| | | CARATS |
|---|---|---|
| From the Kimberley Mine | from | 1·25 to 1·5 |
| ,, De Beers Mine | ,, | 1·20 ,, 1·33 |
| ,, Dutoitspan Mine | ,, | 0·17 ,, 0·5 |
| ,, Bultfontein Mine | ,, | 0·5 ,, 0·33 |

### Varieties of Diamonds
### Fancy Stones

Diamonds occur in all shades, from deep yellow to pure white and jet black, from deep brown to light cinnamon, also green, blue, pink, yellow, orange, and opaque.

Both in Kimberley and De Beers the blue ground on the west side is poorer in diamonds than the blue ground in other parts of the mines. The diamonds from the west side also differ somewhat from those in other parts of the same mine.

The diamonds from each mine have a distinctive character, and so uniform are the characteristics that an experienced buyer can tell at once the locality of any particular parcel of stones. An isolated stone may, of course, be found occasionally in any one mine which is characteristic of some other source of production, but this is the exception to the general rule.

There is a great similarity between the produce of the De Beers and Kimberley mines. A day's wash from either of these mines could be distinguished from each other, but not so easily the majority of the individual stones.

The Kimberley Mine produces a small percentage of white crystals, octahedral in shape, is noted for its large macles, and, in common with the De Beers Mine, it also yields a large percentage of coloured and large yellow diamonds.

The De Beers Mine produces a comparatively small percentage of really white diamonds, but is noted for its fine silvery capes.

The Dutoitspan Mine is noted for its fine white cleavages, silver capes, large yellows, and an exceptional proportion of large stones generally. It also produces a small proportion of fine white, octahedral-shaped crystals and a comparatively small proportion of diamonds below 0·2 of a carat in size.

The Bultfontein Mine produces a very large percentage of white diamonds, mostly octahedral in shape and generally small in size. It produces very few coloured stones, but a larger percentage of flawed and spotted stones than any other mine. Even the apparently pure stones from this mine frequently develop flaws in cutting, which in the rough were imperceptible to the naked eye.

The Wesselton Mine diamonds are noted for an abnormally large percentage of octahedral stones, a large proportion of which are free from flaws. White and brown stones predominate in this mine; there is almost an entire absence of the ordinary yellow, but very fine golden-coloured fancy stones are unearthed occasionally, invariably in the form of cleavage, and hardly ever exceeding 2 carats each in weight.

For "golden fancies" this mine is unrivalled. Wesselton diamonds are easily distinguished from the produce of every other mine by a decided gloss common to them.

Wesselton produces more stones of 10 carats each and over than Bultfontein, but comparatively few large stones of over 50 carats each. It produces a very large percentage of small diamonds under 0·2 of a carat. With Bultfontein it shares the distinction of yielding cubical stones occasionally. It also produces a small percentage of blue-whites.

The Frank Smith Mine produces very fine white diamonds, fairly regular in shape, mostly octahedral, and hardly any coloured stones. Many of the stones are grooved at the edges.

The Kamfersdam Mine yields diamonds of very inferior quality, dark brown being the predominating colour, and even the majority of the better-class stones from this mine are faintly tinged with brown.

The Kimberley West, formerly known as Theron's Mine, situated about 30 miles due west of Kimberley, yields a very small percentage of blue-whites, fine "silver capes," and a large proportion of brown diamonds, somewhat better in quality than Kamfersdam and more regular in shape. The diamonds from this mine present a distinctly "alluvial" appearance, but they are nevertheless distinctive in character from river diamonds and much inferior in quality.

The diamonds from the Leicester Mine are of a distinctive character; they are very much grooved, extremely bad shapes for cutting, and many of the stones are cross-grained.

The Newlands Mine, West Griqualand, about 40 miles north-west of Kimberley, is interesting on account of the occurrence of diamond in what the Reverend Professor Bonney considers to be its true matrix. The workmen occasionally come across well-rounded, boulder-like masses of eclogite, a rather coarsely crystalline rock, sometimes more than a foot in diameter. Some of these boulders have diamonds imbedded in them. One piece examined by Professor Bonney measured approximately 4 inches by 3 inches by 2 inches, and appeared to have been broken off a larger eclogite boulder. In it were seen ten diamonds, mostly well-crystallised octahedra, perfectly colourless, with brilliant lustre, four of them being comprised within a space of a quarter of an inch square. All these diamonds were on the surface. Probably others would have been found inside, but it was not considered desirable to destroy the specimen by breaking it up. It is now in the Natural History Museum, having been presented by the Directors of the Newlands Mine.

Eclogite has been found in other diamond mines, but I am not aware that diamonds have been found imbedded in it except in the Newlands Mine.

Stones from Jagersfontein, in the Orange River Colony, display great purity of colour and brilliancy, and they have the so-called "steely" lustre characteristic of old Indian gems.

### Falling off of Yield with Depth

According to tables furnished by the De Beers Company, the yield of the De Beers and Kimberley mines has declined as the depth increases. At the same time the value of the stones has risen, and diamonds are more expensive to-day than at any previous time.

| YEAR | NUMBER OF CARATS* PER LOAD | VALUE PER CARAT s. | d. |
|---|---|---|---|
| 1889 | 1·283 | 19 | 8·75 |
| 1890 | 1·15 | 32 | 6·75 |
| 1891 | 0·99 | 29 | 6 |
| 1892 | 0·92 | 25 | 6 |
| 1893 | 1·05 | 29 | 0·6 |
| 1894 | 0·89 | 24 | 5·2 |
| 1895 | 0·85 | 25 | 6 |
| 1896 | 0·91 | 26 | 9·4 |
| 1897 | 0·92 | 26 | 10·6 |
| 1898 | 0·80 | 26 | 6·2 |
| 1899 | 0·71 | 29 | 7·2 |
| 1900 | 0·67 | 35 | 10·2 |
| 1901 | 0·76 | 39 | 7 |
| 1902 | 0·76 | 46 | 5·7 |
| 1903 | 0·61 | 48 | 6·3 |
| 1904 | 0·54 | 48 | 11·8 |

* According to Gardner Williams the South African carat is equivalent to 3·174 grains. In Latimer Clark's

## STONES OTHER THAN DIAMONDS

Accompanying diamonds in the concentrates are a number of other minerals of high specific gravity, and some of notable beauty. Among these are the rich red pyrope (garnet), sp. gr. 3·7, containing from

*Dictionary of Metric and other Useful Measures* the diamond carat is given as equal to 3·1683 grains= 0·2053 gramme=4 diamond grains; 1 diamond grain= 0·792 troy grain; 151·5 diamond carats=1 ounce troy.

Webster's *International Dictionary* gives the diamond carat as equal to $3\frac{1}{6}$ troy grains.

*The Oxford English Dictionary* says the carat was originally $\frac{1}{144}$ of an ounce, or $3\frac{1}{3}$ grains, but now equal to about $3\frac{1}{6}$ grains, though varying slightly with time and place.

The *Century Dictionary* says the diamond carat is equal to about $3\frac{1}{6}$ troy grains, and adds that in 1877 the weight of the carat was fixed by a syndicate of London, Paris, and Amsterdam jewellers at 205 milligrammes. This would make the carat equal to 3·163 troy grains. A law has been passed in France ordaining that in the purchase or sale of diamonds and other precious stones the term "metric carat" shall be employed to designate a weight of 200 milligrammes (3·086 grains troy), and prohibiting the use of the word carat to designate any other weight.

1·4 to 3 per cent of oxide of chromium; zircon, in flesh-coloured grains and crystals, sp. gr. 4 to 4·7; kyanite, sp. gr. 3·45 to 3·7, discernible by its blue colour and perfect cleavage; chrome diopside, sp. gr. 3·23 to 3·5, of a bright green colour; bronzite, sp. gr. 3·1 to 3·3; magnetite, sp. gr. 4·9 to 5·2; mixed chrome and titanium iron ore, sp. gr. 4·4 to 4·9, containing from 13 to 61 per cent of oxide of chromium, and from 3 to 68 per cent of titanic acid, in changeable quantities; hornblende, sp. gr. 2·9 to 3·4; barytes, sp. gr. 4·3 to 4·7; and mica. Some of the garnets are of fine quality, and one was recently cut which resembled a pigeon-blood ruby, and attracted an offer of £25.

In the pulsator and sorting house most of the native labourers are long-sentence convicts, supplied with food, clothing, and medical attendance by the Company. They are necessarily well guarded. I myself saw about 1000 convicts at work. I was told that insubordination is very rare; apart from the hopelessness of a successful rising, there is little inducement to revolt ; the lot of these diamond workers is preferable to life in the Government prisons, and they seem contented.

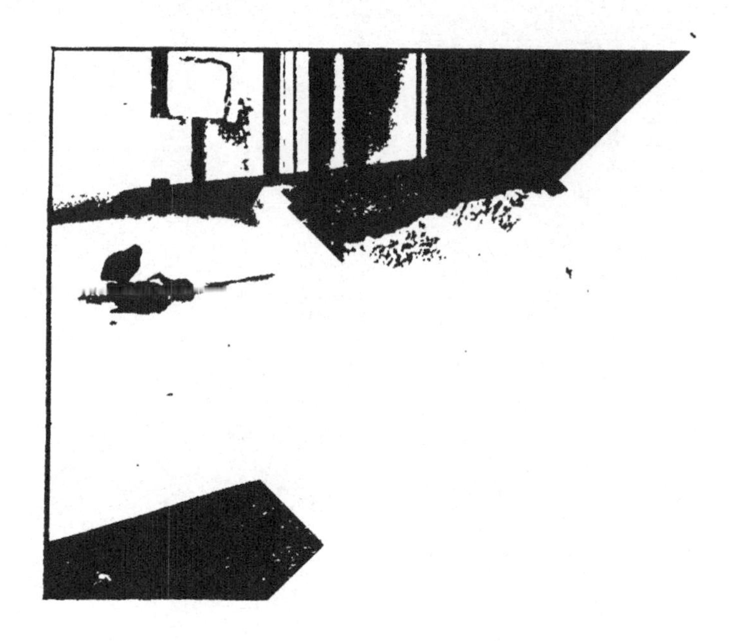

FIG. 10. DE BEERS DIAMOND OFFICE. 25,000 CARATS.

FIG. 11. DE BEERS DIAMOND OFFICE. THE VALUATORS' TABLE.

To face p. 72.

## CHAPTER V

### THE DIAMOND OFFICE

FROM the pulsator the diamonds are sent to the general office in Kimberley to be cleansed in a boiling mixture of nitric and sulphuric acids. A parcel of diamonds loses about half a part per 1000 by this treatment. On one of my visits to the diamond office the door opened and in walked two young men, each carrying a large enamelled saucepan containing something steaming hot. They went to one of the zinc-covered tables and turned out from the saucepans a lustrous heap of 25,000 carats of diamonds (Fig. 10). They had just been boiled in acid and washed.

After purification the diamonds are handed to the valuators (Fig. 11), who sort them into classes, according to size, colour, and purity. In the diamond office they are sorted into ten classes. In the year 1895, in 1141·8 carats of stones, the proportions of the different classes were as follows:

| | |
|---|---|
| Close goods (best stones) | 53·8 |
| Spotted stones | 75·8 |
| Fine cleavage | 79·1 |
| Flats | 39·5 |
| Macles | 36·5 |
| Ordinary and rejection cleavage | 243·4 |
| Rejection stones | 43·2 |
| Light and brown cleavage | 56·9 |
| Rubbish | 371·8 |
| | 1000·0 |
| Fine sand | 141·8 |
| | 1141·8 |

It is a sight for Aladdin to see the valuators at work in the strong-room of the De Beers Company at Kimberley. The tables are literally heaped with stones won from the rough blue ground—stones of all sizes, purified, flashing, and of inestimable price; stones that will be coveted by men and women all the world over; and last, but not least, stones that are probably destined to largely influence the development and history of a whole huge continent.

## CHAPTER VI

### NOTEWORTHY DIAMONDS

PRODIGIOUS diamonds are not so uncommon as is generally supposed. Diamonds weighing over an ounce (151·5 carats) are not unfrequent at Kimberley. Some years ago, in one parcel of stones, I saw eight perfect ounce crystals, and one stone weighing 2 ounces (Fig. 12). The largest diamond from the Kimberley mines weighed 428½ carats, or nearly 4 ounces troy. It measured 1⅞ inch through the longest axis and was 1½ inch square. After cutting it weighed 228½ carats, losing 200 carats in the process. The largest known diamond was discovered in January, 1905, at the New Premier Mine, near Pretoria. This mine is of the same type as the Kimberley mines, but larger in size, and, in fact, is the largest known diamantiferous pipe in the

FIG. 12. A GROUP OF LARGE DIAMOND CRYSTALS.

To face p. 76.

world—the pipe containing the "blue ground," along the longer diameter of its oval-shaped cross-section, measuring over half a mile, and its area is estimated at 350,000 square yards. This pipe breaks through felsitic rocks. The diamond, called "Cullinan" from the name of one of the directors of the company on whose farm it was discovered, was presented to King Edward on his birthday by the people of the Transvaal. It weighed no less than 3025¾ carats, or 9586·5 grains (1·37 lb. avoirdupois). It was a fragment, probably less than half, of a distorted octahedral crystal; the other portions still await discovery by some fortunate miner. The frontispiece shows this diamond in its natural size, from a photograph taken by myself. I had an opportunity of examining and experimenting with this unequalled stone before it was cut. A beam of polarised light passed in any direction through the stone, and then through an analyser, revealed colours in all cases, appearing brightest when the light passed along the greatest diameter—about 4 inches. Here the colours were very fine, but no regular figure was to be seen. Round a small black spot in the interior of the stone the colours were very vivid, changing and rotating round the spot as the analyser was turned. These observations indicated internal strain.

The clearness throughout was remarkable, the stone being absolutely limpid like water, with the exception of a few flaws, dark graphitic spots, and coloured patches close to the outside. At one part near the surface there was an internal crack, showing well the colours of thin plates. At another point there was a milky, opaque mass, of a brown colour, with pieces of what looked like iron oxide. There were four cleavage planes of great smoothness and regularity. On other parts of the surface the crystalline structure was very marked. The edges were rounded in parts, and triangular markings (depressions) were to be seen. I also noticed square depressions, nearly as sharp and perfect as the triangular ones.

The cleaving and cutting and polishing of the Cullinan diamond was entrusted to the firm of Asscher and Co., in Amsterdam. The cleavage of the diamond was very successfully accomplished by Mr. Joseph Asscher. An incision half an inch deep was made with a sharp diamond point in the proper place, then a specially designed knife blade was placed in the incision and it was struck a heavy blow with a piece of steel. The diamond split through a defective spot, part of which was left in each portion of the diamond.

Gigantic as is the Cullinan diamond, it represents in weight less than half the daily output of the De Beers mines, which averages about 7000 carats per day.

Next in size to the Cullinan comes the one which was found at the Jagersfontein Mine. It weighed 970 carats—over half a pound.

## DIAMONDS

The following table gives the names and weights of some historic diamonds (Fig. 13):

1. Koh-i-noor, after the second cutting, 106 carats.
2. Loterie d'Angleterre, 49 carats.
3. Nizam of Hyderabad, 279 carats.
4. Orloff, 194 carats.
5. Koh-i-noor, after first cutting, 279 carats.
6. Regent or Pitt, 137 carats.
7. Duke of Tuscany, 133 carats.
8. Star of the South, 124 carats.
9. Pole Star, 40 carats.
10. Tiffany, yellow, 125 carats.
11. Hope, blue diamond, 44 carats.
12. Sancy, 53 carats.
13. Empress Eugenie, 51 carats.
14. Shah, 86 carats.
15. Nassak, 79 carats.
16. Pasha of Egypt, 40 carats.
17. Cullinan, 3025 carats.
18. Excelsior, Jagersfontein, 969 carats.

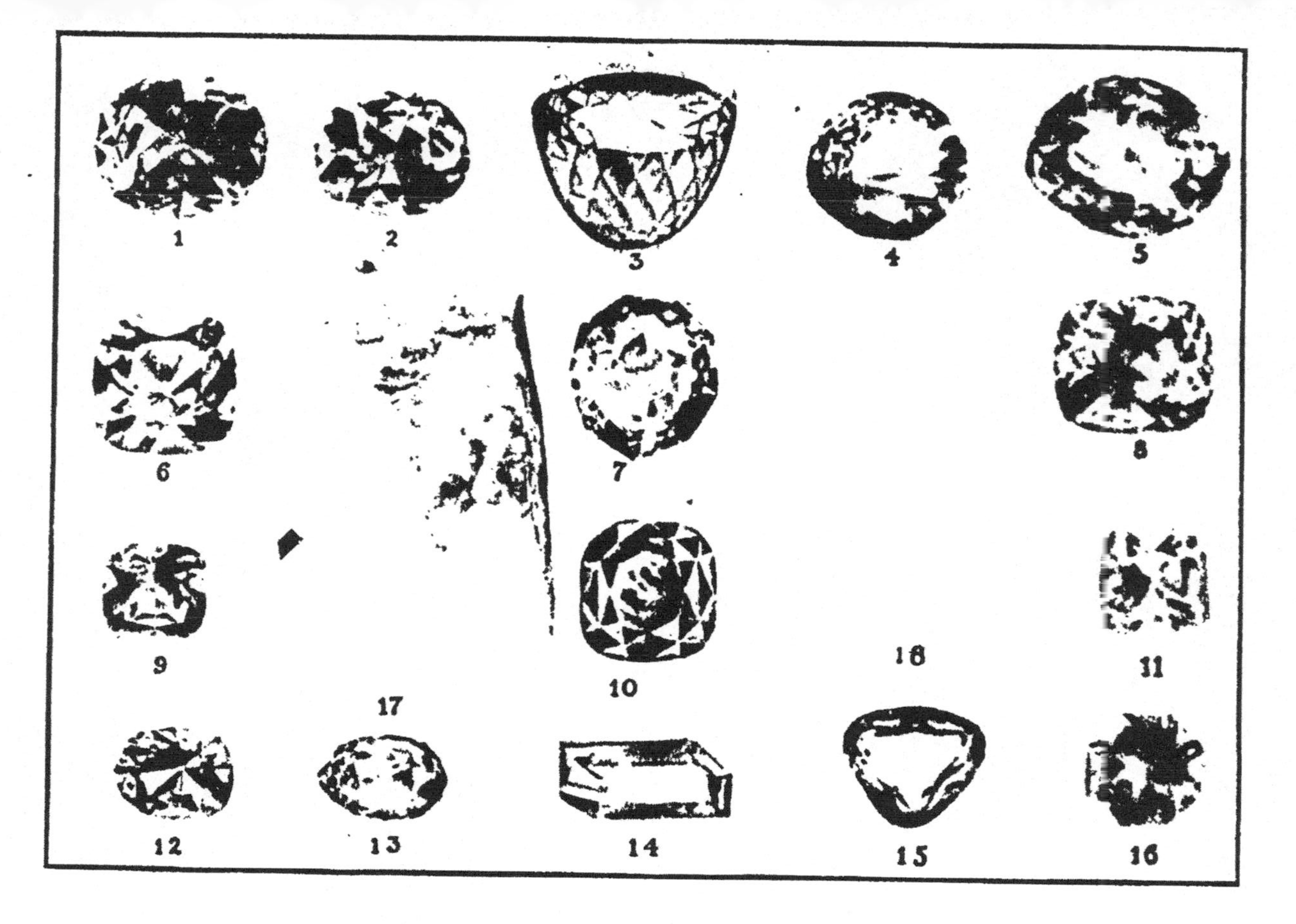


FIG. 13. SOME HISTORIC DIAMONDS.

To face p. 80.

## CHAPTER VII

### BOART, CARBONADO, AND GRAPHITE

THE black inclusions in some transparent diamonds consist of graphite. On crushing a clear diamond showing such spots and heating in oxygen to a temperature well below the point at which diamond begins to burn, Moissan found that the grey tint of the powder disappeared, no black spots being seen under the microscope. There also occur what may be considered intermediate forms between the well-crystallised diamond and graphite. These are "boart" and "carbonado." Boart is an imperfectly crystallised diamond, having no clear portions, and therefore useless for gems. Shot boart is frequently found in spherical globules, and may be of all colours. Ordinary boart is so hard that it is used in rock-drilling, and when crushed it is employed for cutting and polishing other stones. Carbonado is the Brazilian term for a still less perfectly crystallised form of carbon. It is equally hard, and occurs in porous masses and in massive black pebbles, sometimes weighing two or more ounces.

The ash left after burning a diamond invariably contains iron as its chief constituent; and the most common colours of diamonds, when not perfectly pellucid, show various shades of brown and yellow, from the palest "off colour" to almost black. These variations give support to the theory advanced by Moissan that the diamond has separated from molten iron—a theory of which I shall say more presently—and also explain how it happens that stones from different mines, and even from different parts of the same mine, differ from each other. Further confirmation is given by the fact that the country round Kimberley is remarkable for its ferruginous character, and iron-saturated soil is popularly regarded as one of the indications of the near presence of diamonds.

## Graphite

Intermediate between soft carbon and diamond come the graphites. The name graphite is given to a variety of carbon, generally crystalline, which in an oxidising mixture of chlorate of potassium and nitric acid forms graphitic oxide. This varies in colour from green to brown or yellow, or it is almost without colour, according to the completeness of the reaction. Graphites are of varying densities, from 2·0 to 3·0, and generally of crystalline aspect. Graphite and diamond pass insensibly into one another. Hard graphite and soft diamond are near the same specific gravity. The difference appears to be one of pressure at the time of formation.

Some forms of graphite exhibit the remarkable property by which it is possible to ascertain approximately the temperature at which they were formed, or to which they have subsequently been exposed. Sprouting graphite is a form, frequently met with in nature, which on moderate heating swells up to a bulky, very light mass of amorphous carbon. Moissan has found it in blue ground from Kimberley; my own results verify his. When obtained by simple elevation of temperature in the arc or the electric furnace graphites do not sprout; but when they are formed by dissolving carbon in a metal at a high temperature and then allowing the graphite to separate out on cooling, the sprouting variety appears. The phenomenon of sprouting is easily shown. If a few grains are placed in a test-tube and heated to about 170° C., the grains increase enormously in bulk and fill the tube with a light form of amorphous carbon.

The resistance of a graphite to oxidising agents is greater the higher the temperature to which it has previously been exposed. Graphites which are easily attacked by a mixture of fuming nitric acid and potassium chlorate are rendered more resistant by strong heat in the electric furnace.

I have already signified that there are various degrees of refractoriness to chemical reagents among the different forms of graphite. Some dissolve in strong nitric acid; other forms of graphite require a mixture of highly concentrated nitric acid and potassium chlorate to attack them, and even with this intensely powerful agent some graphites resist longer than others. M. Moissan has shown that the power of resistance to nitric acid and potassium chlorate is in proportion to the temperature at which the graphite was formed, and with tolerable certainty we can estimate this temperature by the resistance of the specimen of graphite to this reagent.

## Crystallisation

The diamond belongs to the isometric system of crystallography; the prevailing form is octahedral. It frequently occurs with curved faces and edges. Twin crystals (macles) are not uncommon. Diamond crystals are generally perfect on all sides. They seldom show irregular sides or faces by which they were attached to a support, as do artificial crystals of chemical salts; another proof that the diamond must have crystallised from a dense liquid.

The accompanying illustration (Fig. 14) shows some of the various crystalline forms of native diamonds.

No. 1. Diamond in the form of a hexakis-octahedron (the forty-eight scalenohedron), or a solid figure contained by forty-eight scalene triangles. According to Professor Maskelyne, this occurs as a self-existent form only in the diamond.

No. 2. Diamond in the form of a hexakis-

7.

4.

1.

5.

2.

6.

3.

FIG. 14. CRYSTALLINE FORMS OF NATIVE DIAMONDS.

To face p. 86.

octahedron and octahedron. From Sudafrika.

No. 3. Diamond in the form of octahedron with intersections.

No. 4. Diamond from Brazil.

No. 5. Diamond from Kimberley.

No. 6. Diamond from Brazil.

No. 7. A macle or twin crystal, showing its formation from an octahedron with curved edges.

Some crystals of diamonds have their surfaces beautifully marked with equilateral triangles, interlaced and of varying sizes (Fig. 15). Under the microscope these markings appear as hollow depressions sharply cut out of the surrounding surface, and these depressions were supposed by Gustav Rose to indicate the probability that the diamonds had at some previous time been exposed to incipient combustion. Rose pointed out that similar triangular striations appeared on the surfaces of diamonds burnt before the blowpipe. This experiment I have repeated on a clear diamond, and I have satisfied myself that during combustion before the blowpipe, in the field of a microscope, the surface is etched with triangular markings different in character from those naturally on crystals (Fig. 16). The artificial striæ are very irregular, much smaller, and massed closer together, looking as if the diamond during combustion flaked away in triangular chips, while the markings natural to crystals appear as if produced by the crystallising force as they were being built up. Many crystals of chemical compounds appear striated from both these causes. Geometrical markings can be produced by eroding the surface of a crystal of alum with water, and they also occur naturally during crystallisation.

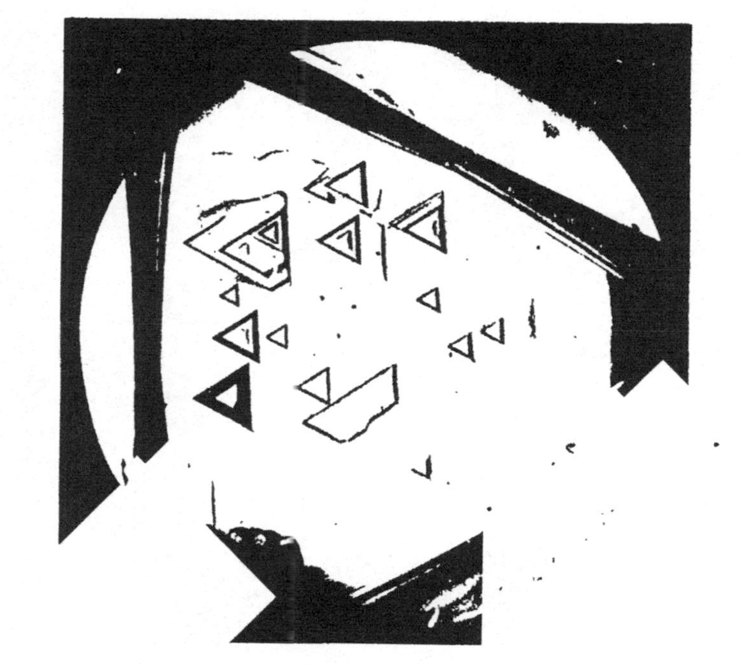

FIG. 15. TRIANGULAR MARKINGS ON NATURAL FACE OF A DIAMOND CRYSTAL.

FIG. 16. TRIANGULAR MARKINGS ARTIFICIALLY PRODUCED ON A DIAMOND CRYSTAL.

To face page 88

## CHAPTER VIII

### PHYSICAL AND CHEMICAL PROPERTIES OF THE DIAMOND

I NEED scarcely say the diamond is almost pure carbon, and it is the hardest substance in nature.

When heated in air or oxygen to a temperature varying from 760° to 875° C., according to its hardness, the diamond burns with production of carbonic acid. It leaves an extremely light ash, sometimes retaining the shape of the crystal, consisting of iron, lime, magnesia, silica, and titanium. In boart and carbonado the amount of ash sometimes rises to 4 per cent, but in clear crystallised diamonds it is seldom higher than 0·05 per cent. By far the largest constituent of the ash is iron.

The following table shows the temperatures of combustion in oxygen of different kinds of carbon:

| | ° C. |
|---|---|
| Condensed vapour of carbon | 650 |
| Carbon from sugar, heated in an electrical furnace | 660 |
| Artificial graphites, generally | 660 |
| Graphite from ordinary cast-iron | 670 |
| Carbon from blue ground, of an ochre colour | 690 |
| Carbon from blue ground, very hard and black | 710 |
| Diamond, soft Brazilian | 760 |
| Diamond, hard Kimberley | 780 |
| Boart from Brazil | 790 |
| Boart from Kimberley | 790 |
| Boart, very hard, almost impossible to cut | 900 |

### Hardness

Diamonds vary considerably in hardness, and even different parts of the same crystal differ in their resistance to cutting and grinding.

## HARDNESS OF DIAMOND

Beautifully white diamonds have been found at Inverel, New South Wales, and from the rich yield of the mine and the white colour of the stones great things were expected. In the first parcel which came to England the stones were found to be so much harder than South African diamonds that it was at first feared they would be useless except for rock-boring purposes. The difficulty of cutting them disappeared with improved appliances, and they now are highly prized.

The famous Koh-i-noor, when being cut into its present form, showed a notable variation in hardness. In cutting one of the facets near a yellow flaw, the crystal became harder and harder the further it was cut, until, after working the mill for six hours at the usual speed of 2400 revolutions a minute, little impression was made. The speed was increased to more than 3000, when the work slowly proceeded. Other portions of the stone were found to be comparatively soft, and became harder as the outside was cut away.

The intense hardness of the diamond can be illustrated by the following experiment. On the flattened apex of a conical block of steel place a diamond, and upon it bring down a second cone of steel. On forcing together the two steel cones by hydraulic pressure the stone is squeezed into the steel blocks without injuring it in the slightest degree.

In an experiment I made at Kimberley the pressure gauge showed 60 atmospheres, and the piston being 3·2 inches diameter, the absolute pressure was 3·16 tons, equivalent on a diamond of 12 square mm. surface to 170 tons per square inch of diamond.

The use of diamond in glass-cutting I need not dwell on. So hard is diamond in comparison to glass, that a suitable splinter of diamond will plane curls off a glass plate as a carpenter's tool will plane shavings off a deal board. The illustration (Fig. 17) shows a few diamond-cut glass shavings.

## Density or Specific Gravity

The specific gravity of the diamond varies ordinarily from 3·514 to 3·518. For comparison, I give in tabular form the specific gravities of the different varieties of carbon and of the minerals found on the sorting tables :

| | SPECIFIC GRAVITY. |
|---|---|
| Amorphous carbon | 1·45–1·70 |
| Hard gas coke | 2·356 |
| Hard graphite | 2·5 |
| Quartzite and granite | 2·6 |
| Beryl | 2·7 |
| Mica | 2·8 |
| Hornblende | 3·0 |
| Boart | 3·47–3·49 |
| Carbonado | 3·50 |
| Diamond | 3·514–3·518 |
| Garnet | 3·7 |
| Corundum | 3·8 |
| Zircon | 4·4 |
| Barytes | 4·5 |
| Chrome and titanic iron ore | 4·7 |
| Magnetite | 5·0 |

# DIAMONDS

There is a substance, the double nitrate of silver and thallium, which, while solid at ordinary temperatures, liquefies at 75° C. and then has a specific gravity of 4·5. Admixture with water lowers the density to any desired point.

If a glass cell is taken containing this liquid diluted to a density of about 3·6, and in it is thrown pieces of the above-named minerals, all those whose density is lower than 3·6 will rise to the surface, while the denser minerals will sink. If now a little water is carefully added with constantly stirring until the density of the liquid is reduced to that of the diamond, the heterogeneous collection sorts itself into three parts. The graphite, quartz, beryl, mica, and hornblende rise to the surface; the garnet, corundum, zircons, etc., sink to the bottom, while the diamonds float in the middle of the liquid. With a platinum landing-net I can skim off the swimmers and put them into one dish; with the same net I can fish out the diamonds and put them in a second dish, while by raising a sieve at the bottom I can remove the heavy minerals and put them into a third. The accurate separation of diamonds from the heterogeneous mixture can be effected in less time than is taken to describe the experiment.

The table shows that diamonds vary somewhat in density among themselves, between narrow limits. Occasionally, however, diamonds overpass these figures. Here is an illustration. In a test-tube of the same dense liquid are three selected diamonds. One rises to the top, another floats uncertain where to settle, rising and falling as the temperature of the sorting liquid is raised or lowered, whilst the third sinks to the bottom. Allowing the liquid to cool a degree or two slightly increases the density and sends all three to the surface.

## Phosphorescence of Diamond

After exposure for some time to the sun many diamonds glow in a dark room. Some diamonds are fluorescent, appearing milky in sunlight. In a vacuum, exposed to a high-tension current of electricity, diamonds phosphoresce of different colours, most South African diamonds shining with a bluish light. Diamonds from other localities emit bright blue, apricot, pale blue, red, yellowish green, orange, and pale green light. The most phosphorescent diamonds are those which are fluorescent in the sun. One beautiful green diamond in the writer's collection, when phosphorescing in a good vacuum, gives almost as much light as a candle, and you can easily read by its rays. But the time has hardly come when diamonds can be used as domestic illuminants! The emitted light is pale green, tending to white, and in its spectrum, when strong, can be seen bright lines, one at about λ 5370 in the green, one at λ 5130 in the greenish blue, and one at λ 5030 in the blue. A beautiful collection of diamond crystals belonging to Professor Maskelyne phosphoresces with nearly all the colours of the rainbow, the different faces glowing with different shades of colour. Diamonds which phosphoresce red generally show the yellow sodium line on a continuous spectrum. In one Brazilian diamond phosphorescing a reddish-yellow colour I detected in its spectrum the citron line characteristic of yttrium.

The rays which make the diamond phosphoresce are high in the ultra-violet. To illustrate this phosphorescence under the influence of the ultra-violet rays, arrange a powerful source of these rays, and in front expose a design made up of certain minerals, willemite, franklinite, calcite, etc.—phosphorescing of different colours. Their brilliant glow ceases entirely when a thin piece of glass is interposed between them and the ultra-violet lamp.

I now draw attention to a strange property of the diamond, which at first sight might seem to discount the great permanence and unalterability of this stone. It has been ascertained that the cause of phosphorescence is in some way connected with the hammering of the electrons, violently driven from the negative pole on to the surface of the body under examination, and so great is the energy of the bombardment, that impinging on a piece of platinum or even iridium, the metal will actually melt. When the diamond is thus bombarded in a radiant matter tube the result is startling. It not only phosphoresces, but becomes discoloured, and in course of time becomes black on the surface. Some diamonds blacken in the course of a few minutes, while others require an hour or more to discolour. This blackening is only superficial, and although no ordinary means of cleaning will remove the discolouration, it goes at once when the stone is polished with diamond powder.

FIG. 17. DIAMOND-CUT GLASS AND SHAVINGS.

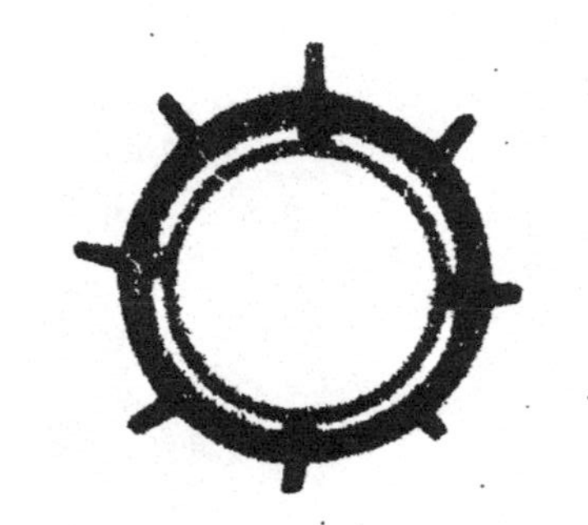


FIG. 18. DIAMONDS IN RÖNTGEN RAYS.

A. BLACK DIAMOND IN GOLD FRAME.
B. PINK DELHI DIAMOND.
C. PASTE IMITATION OF B.

To face p. 98.

Ordinary oxidising reagents have little or no effect in restoring the colour.

The superficial dark coating on a diamond after exposure to molecular bombardment I have proved to be graphite. M. Moissan has shown that this graphite, on account of its great resistance to oxidising reagents, cannot have been formed at a lower temperature than 3600° C.

It is thus manifest that the bombarding electrons, striking the diamond with enormous velocity, raise the superficial layer to the temperature of the electric arc and turn it into graphite, whilst the mass of diamond and its conductivity to heat are sufficient to keep down the general temperature to such a point that the tube appears scarcely more than warm to the touch.

A similar action occurs with silver, the superficial layers of which can be raised to a red heat without the whole mass becoming more than warm.

## Conversion of Diamond into Graphite

Although we cannot convert graphite into diamond, we can change the diamond into graphite. A clear crystal of diamond is placed between two carbon poles, and the poles with intervening diamond are brought together and an arc formed between. The temperature of the diamond rapidly rises, and when it approaches 3600° C., the vaporising point of carbon, it breaks down, swells, and changes into black and valueless graphite.

## Tribo-Luminescence

A few minerals give out light when rubbed. In the year 1663 the Hon. Robert Boyle read a paper before the Royal Society, in which he described several experiments made with a diamond which markedly showed tribo-luminescence. As specimens of tribo-luminescent bodies I may instance sphalerite (sulphide of zinc), and an artificial sphalerite, which is even more responsive to friction than the native sulphide.*

Mrs. Kunz, wife of the well-known New York mineralogist, possesses, perhaps, the most remarkable of all phosphorescing diamonds. This prodigy diamond will phosphoresce in the dark for some minutes after being exposed to a small pocket electric light, and if rubbed on a piece of cloth a long streak of phosphorescence appears.

## Absorption Spectrum of Diamond

On passing a ray of light through a diamond and examining it in a spectroscope, Walter has found in all colourless brilliants of over 1 carat in weight an absorption band at wave-length 4155 (violet). He

* Artificial tribo-luminescent sphalerite:—

| | |
|---|---|
| Zinc carbonate . . . | 100 parts |
| Flower of sulphur . . | 30 ,, |
| Manganese sulphate . . | ½ per cent. |

Mix with distilled water and dry at a gentle heat. Put in luted crucible and keep at a bright red heat for from two to three hours.

ascribes this band to an impurity and suggests it may possibly be due to samarium. Three other fainter lines were detected in the ultra-violet by means of photography.

### Refractivity

But it is not the hardness of the diamond so much as its optical qualities that make it so highly prized. It is one of the most refracting substances in nature, and it also has the highest reflecting properties. In the cutting of diamonds advantage is taken of these qualities. When cut as a brilliant the facets on the lower side are inclined so that light falls on them at an angle of 24° 13′, at which angle all the incident light is totally reflected. A well-cut brilliant should appear opaque by transmitted light except at a small spot in the middle where the table and culet are opposite. All the light falling on the front of the stone is reflected from the facets, and the light passing into the diamond is reflected from the interior surfaces and refracted into colours when it passes out into the air, giving rise to the lightnings, the effulgence, and coruscations for which the diamond is supreme above all other gems.

The following table gives the refractive indices of diamonds and other bodies :

| Refractive Indices for the D Line | |
|---|---|
| Chromate of lead | 2·50–2·97 |
| Diamond | 2·47–2·75 |
| Phosphorus | 2·22 |
| Sulphur | 2·12 |
| Ruby | 1·78 |
| Thallium glass | 1·75 |
| Iceland spar | 1·65 |
| Topaz | 1·61 |
| Beryl | 1·60 |
| Emerald | 1·59 |
| Flint glass | 1·58 |
| Quartz | 1·55 |
| Canada balsam | 1·53 |
| Crown glass | 1·53 |
| Fluor-spar | 1·44 |
| Ice | 1·31 |

In vain I have searched for a liquid of the same refraction as diamond. Such a liquid would be invaluable to the merchant, as on immersing a stone the clear body would absolutely disappear, leaving in all their ugliness the flaws and black specks so frequently seen even in the best stones.

### The Diamond and Polarised Light

Having no double refraction, the diamond should not act on polarised light. But as is well known, if a transparent body which does not so act is submitted to strain of an irregular character it becomes doubly refracting, and in the polariscope reveals the existence of the strain by brilliant colours arranged in a more or less defined pattern, according to the state of tension in which the crystal exists. I have examined many hundred diamond crystals under polarised light, and with few exceptions the colours show how great is the strain to which some of them are exposed. On rotating the polariser, the black cross most frequently seen revolves round a particular point in the inside of the crystal; on examining this point with a high power we sometimes see a slight flaw, more rarely a minute cavity. The cavity is filled with gas at enormous pressure, and the strain is set up in the stone by the effort of the gas to escape. I have already said that the great Cullinan diamond by this means revealed a state of considerable internal stress and strain.

So great is this strain of internal tension that it is not uncommon for a diamond to explode soon after it reaches the surface, and some have been known to burst in the pockets of the miners or when held in the warm hand. Large crystals are more liable to burst than smaller pieces. Valuable stones have been destroyed in this way, and it is whispered that cunning dealers are not averse to allowing responsible clients to handle or carry in their warm pockets large crystals fresh from the mine. By way of safeguard against explosion some dealers imbed large diamonds in raw potato to ensure safe transit to England.

The anomalous action which many diamonds exert on polarised light is not such as can be induced by heat, but it can easily be conferred on diamonds by pressure, showing that the strain has not been produced by sudden cooling, but by sudden lowering of pressure.

The illustration of this peculiarity is not only difficult, but sometimes exceedingly costly—difficult because it is necessary to arrange for projecting on the screen the image of a diamond crystal between the jaws of a hydraulic press, the illuminating light having to pass through delicate optical polarising apparatus—and costly because only perfectly clear crystals can be used, and crystals of this character sometimes fly to pieces as the pressure rises. At first no colour is seen on the screen, the crystal not being birefringent. A movement of the handle of the press, however, gives the crystal a pinch, instantly responded to by the colours on the screen, showing the production of double refraction. Another movement of the handle brightens the colours, and a third may strain the crystal beyond its power of resistance, when the crystal flies to pieces.

### The Diamond and Röntgen Rays

The diamond is remarkable in another respect. It is extremely transparent to the Röntgen rays, whereas highly refracting glass, used in imitation diamonds, is almost perfectly opaque to the rays. I exposed for a few seconds over a photographic plate to the X-rays the large Delhi diamond of a rose-pink colour weighing 31½ carats, a black diamond weighing 23 carats, and a glass imitation of the pink diamond (Fig. 18). On development the impression where the diamond obscured the rays was found to be strong, showing that most rays passed through, while the glass was practically opaque. By this means imitation diamonds can readily be distinguished from true gems.

### Action of Radium on Diamond

The $\beta$-rays from radium having like properties to the stream of negative electrons in a radiant matter tube, it was of interest to ascertain if they would exert a like difference on diamond. The diamond glows under the influence of the $\beta$-radiations, and crushed diamond cemented to a piece of card or metal makes an excellent screen in a spinthariscope—almost as good as zinc sulphide. Some colourless crystals of diamond were imbedded in radium bromide and kept undisturbed for more than twelve months. At the end of that time they were examined. The radium had caused them to assume a bluish-green colour, and their value as " fancy stones " had been increased.

This colour is persistent and penetrates

below the surface. It is unaffected by long-continued heating in strong nitric acid and potassium chlorate, and is not discharged by heating to redness.

To find out if this prolonged contact with radium had communicated to the diamond any radio-active properties, six diamonds were put on a photographic plate and kept in the dark for a few hours. All showed radio-activity by darkening the sensitive plate, some being more active than others. Like the green tint, the radio-activity persists after drastic treatment. To me this proves that radio-activity does not merely consist in the adhesion of electrons or emanations given off by radium to the surface of an adjacent body, but the property is one involving layers below the surface, and like the alteration of tint, is probably closely connected with the intense molecular excitement the stone had experienced during its twelve months' burial in radium bromide.

A diamond that had been coloured by

radium, and had acquired strong radio-active properties, was slowly heated to dull redness in a dark room. Just before visibility a faint phosphorescence spread over the stone. On cooling and examining the diamond it was found that neither the colour nor the radio-activity had suffered appreciably.

### Boiling- and Melting-point of Carbon

On the average the critical point of a substance is 1·5 times its absolute boiling-point. Therefore the critical point of carbon should be about 5800° Ab. But the absolute critical temperature divided by the critical pressure is for all the elements so far examined never less than 2·5; this being about the value Sir James Dewar finds for hydrogen. So that, accepting this, we get the maximum critical pressure as follows, viz. 2320 atmospheres:

$$\frac{5800^\circ \text{ Ab.}}{\text{CrP}} = 2.5, \text{ or CrP} = \frac{5800 \text{ Ab.}}{2.5},$$

or 2320 atmospheres.

# MELTING-POINT OF CARBON

Carbon and arsenic are the only two elements that have a melting-point above the boiling-point; and among compounds carbonic acid and fluoride of silicium are the only other bodies with similar properties. Now the melting-point of arsenic is about 1·2 times its absolute boiling-point. With carbonic acid and fluoride of silicium the melting-points are about 1·1 times their boiling-points. Applying these ratios to carbon, we find that its melting-point would be about 4400°.

Therefore, assuming the following data:

Boiling-point ..............3870° Ab.
Melting-point..............4400°
Critical temperature .......5800°
Critical pressure ...........2320 Ats.

the Rankine or Van der Waals formula, calculated from the boiling-point and critical data, would be as follows:

$$\log. P = 10{\cdot}11 - 39120/T,$$

and this gives for a temperature of 4400° Ab. a pressure of 16·6 Ats. as the melting-point pressure. The results of the formula are given in the form of a table :

| Temperature Ab. | Pressure Ats. | |
|---|---|---|
| 3870° | 1·00 | Boiling-point. |
| 4000° | 2·14 | |
| 4200° | 6·25 | |
| 4400° | 16·6 | Melting-point. |
| 4600° | 40·4 | |
| 4800° | 91·2 | |
| 5000° | 193 | |
| 5200° | 386 | |
| 5400° | 735 | |
| 5600° | 1330 | |
| 5800° | 2320 | Critical point (15 tons per square inch). |

If, then, we may reason from these rough estimates, above a temperature of 5800° Ab. no amount of pressure will cause carbon vapour to assume liquid form, whilst at 4400° Ab. a pressure of above 17 atmospheres would suffice to liquefy some of it. Between these extremes the curve of vapour pressure is assumed to be logarithmic, as represented in the accompanying diagram.

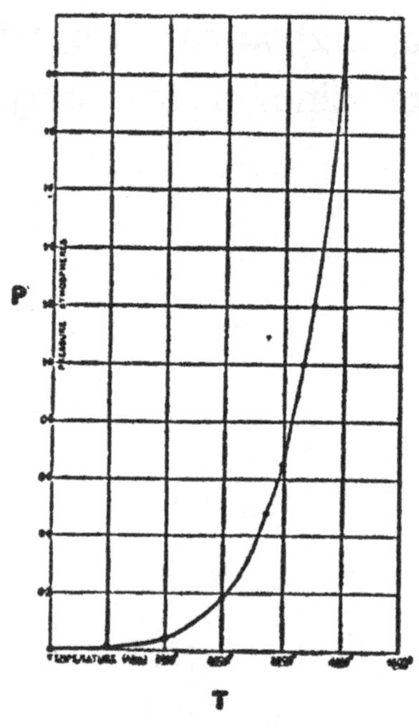


FIG. 19. CURVE OF VAPOUR PRESSURE OF CARBON

The constant 39120 which occurs in the logarithmic formula enables us to calculate the latent heat of evaporation. If we assume the vapour density to be normal, or the molecule in vapour as $C_2$, then the heat of volatilisation of 12 grms. of carbon would be 90,000 calories; or, if the vapour is a condensed molecule like $C_6$, then the 12 grms. would need 30,000 calories. In the latter case the evaporation of 1 grm. of carbon would require 2500 calories, whereas a substance like zinc needs only about 400 calories.

## CHAPTER IX

### GENESIS OF THE DIAMOND

SPECULATIONS as to the probable origin of the diamond have been greatly forwarded by patient research, and particularly by improved means of obtaining high temperatures, an advance we owe principally to the researches of the late Professor Moissan.

Until recent years carbon was considered absolutely non-volatile and infusible; but the enormous temperatures placed at the disposal of experimentalists by the introduction of electricity show that, instead of breaking rules, carbon obeys the same laws that govern other bodies. It volatilises at the ordinary pressure at a temperature of about 3600° C., and passes from the solid to the gaseous state without liquefying. It has been found that other bodies, such as arsenic, which volatilise without liquefying at the ordinary pressure, will easily liquefy if pressure is added to temperature. It naturally follows that if along with the requisite temperature sufficient pressure is applied, liquefaction of carbon will take place, when on cooling it will crystallise. But carbon at high temperatures is a most energetic chemical agent, and if it can get hold of oxygen from the atmosphere or any compound containing it, it will oxidise and fly off in the form of carbonic acid. Heat and pressure therefore are of no avail unless the carbon can be kept inert.

It has long been known that iron, when melted, dissolves carbon, and on cooling liberates it in the form of graphite. Moissan discovered that several other metals, especially silver, have similar properties; but iron is the best solvent for carbon. The quantity of carbon entering into solution increases with the temperature.

For the artificial manufacture of diamond the first necessity is to select pure iron—free

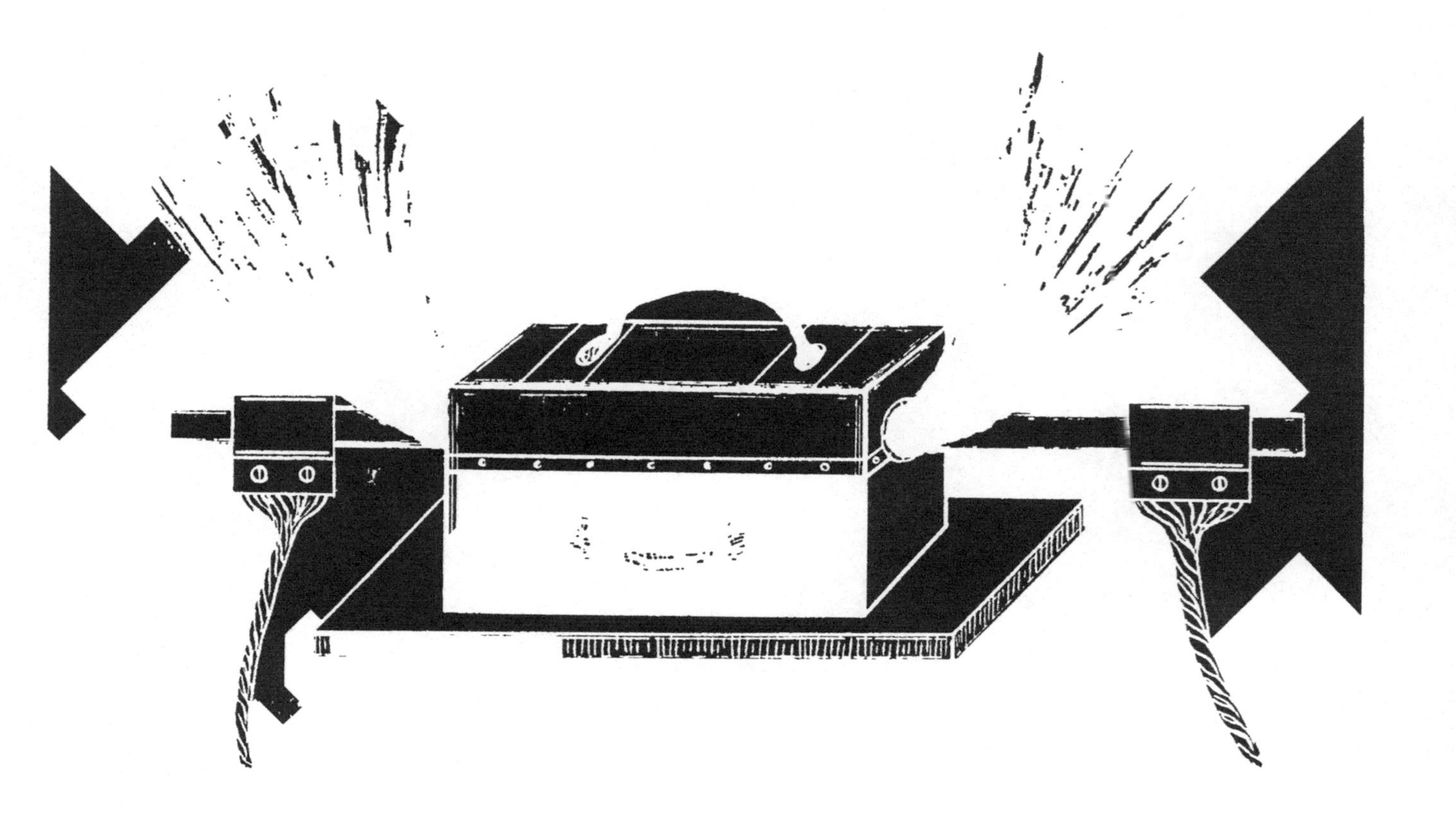

FIG. 20. MOISSAN'S ELECTRIC FURNACE.

To face p. 116.

from sulphur, silicon, phosphorus, etc.—and to pack it in a carbon crucible with pure charcoal from sugar. The crucible is then put into the body of the electric furnace and a powerful arc formed close above it between carbon poles, utilising a current of 700 ampères at 40 volts pressure (Fig. 20). The iron rapidly melts and saturates itself with carbon. After a few minutes' heating to a temperature above 4000° C.—a temperature at which the iron melts like wax and volatilises in clouds—the current is stopped and the dazzling fiery crucible is plunged beneath the surface of cold water, where it is held till it sinks below a red heat. As is well known, iron increases in volume at the moment of passing from the liquid to the solid state. The sudden cooling solidifies the outer layer of iron and holds the inner molten mass in a tight grip. The expansion of the inner liquid on solidifying produces an enormous pressure, and under the stress of this pressure the dissolved carbon separates out in transparent forms—minutely microscopic, it is true—all the same veritable diamonds, with crystalline form and appearance, colour, hardness, and action on light, the same as the natural gem.

Now commences the tedious part of the process. The metallic ingot is attacked with hot nitro-hydrochloric acid until no more iron is dissolved. The bulky residue consists chiefly of graphite, together with translucent chestnut-coloured flakes of carbon, black opaque carbon of a density of from 3·0 to 3·5 and hard as diamonds—black diamonds or carbonado, in fact—and a small portion of transparent, colourless diamonds showing crystalline structure. Besides these there may be carbide of silicon and corundum, arising from impurities in the materials employed.

The residue is first heated for some hours with strong sulphuric acid at the boiling-point, with the cautious addition of powdered nitre. It is then well washed and for two days allowed to soak in strong hydrofluoric acid in cold, then in boiling acid. After this treatment the soft graphite disappears, and most, if not all, the silicon compounds have been destroyed. Hot sulphuric acid is again applied to destroy the fluorides, and the residue, well washed, is attacked with a mixture of the strongest nitric acid and powdered potassium chlorate, kept warm—but not above 60° C., to avoid explosions. This treatment must be repeated six or eight times, when all the hard graphite will gradually be dissolved and little else left but graphitic oxide, diamond, and the harder carbonado and boart. The residue is fused for an hour in fluorhydrate or fluoride of potassium, then boiled out in water and again heated in sulphuric acid. The well-washed grains which resist this energetic treatment are dried, carefully deposited on a slide, and examined under the microscope. Along with numerous pieces of black diamond are seen transparent, colourless pieces, some amorphous, others with a crystalline appearance. Fig. 21 B shows one of these crystalline fragments. Although many fragments of crystals occur, it is remarkable I have never seen a complete crystal. All appear shattered, as if on being liberated from the intense pressure under which they were formed they burst asunder. I have singular evidence of this phenomenon. A fine piece of artificial diamond, carefully mounted by me on a microscopic slide, exploded during the night and covered the slide with fragments. Moissan's crystals of artificial diamond sometimes broke a few weeks after their preparation, and some of the diamonds which cracked weeks or even months after their preparation showed fissures covered with minute cubes. I have explained that this bursting paroxysm is not unknown at the Kimberley mines. So far, all such artificial diamonds are microscopic. The largest artificial diamond is less than one millimetre across.

These laboratory diamonds burn in the air before the blowpipe to carbonic acid. In lustre, crystalline form, optical properties, density, and hardness they are identical with the natural stone.

In several cases Moissan separated ten to fifteen microscopic diamonds from a single ingot. The larger of these are about 0·75 mm. long, the octahedra being 0·2 mm.

The accompanying illustrations (Fig. 22) are copied from drawings in Moissan's book *Le Four Electrique*.

Along with carbon, molten iron dissolves other bodies which possess tinctorial powers. We know of blue, green, pink, yellow, and orange diamonds. One batch of iron might contain an impurity colouring the stones blue, another lot would tend towards the formation of pink stones, another of green, and so on. Cobalt, nickel, chromium, and manganese, all metals present in the blue ground, would produce these colours.

## A New Formation of Diamond

I have long speculated as to the possibility of obtaining artificially such pressures and temperatures as would fulfil the above conditions. In their researches on the gases from fired gunpowder and cordite, Sir Frederick Abel and Sir Andrew Noble obtained in closed steel cylinders pressures as great as 95 tons to the square inch, and temperatures as high as 4000° C. According to a paper recently communicated to the Royal Society, Sir Andrew Noble, exploding cordite in closed vessels, has obtained a pressure of 8000 atmospheres, or 50 tons per square inch, with a temperature reaching in all probability 5400° Ab.

Here, then, we have conditions favourable for the liquefaction of carbon, and were the time of explosion sufficient to allow the reactions to take place, we should certainly expect to get the liquid carbon to solidify in the crystalline state.*

By the kindness of Sir Andrew Noble I have been enabled to work upon some of the residues obtained in closed vessels after explosions, and I have submitted them to the same treatment that the granulated iron had gone through. After weeks of patient toil I removed the amorphous carbon, the graphite, the silica,† and other constituents

* Sir James Dewar, in a Friday evening discourse at the Royal Institution in 1880, showed an experiment proving that the temperature of the interior of a carbon tube heated by an outside electric arc was higher than that of the oxy-hydrogen flame. He placed a few small crystals of diamond in the carbon tube, and, maintaining a current of hydrogen to prevent oxidation, raised the temperature of the tube in an electric furnace to that of the arc. In a few minutes the diamond was transformed into graphite. At first sight this would seem to show that diamond cannot be formed at temperatures above that of the arc. It is probable, however, for reasons given above, that at exceedingly high pressures the result would be different.

† The silica was in the form of spheres, perfectly shaped and transparent, mostly colourless, but among them several of a ruby colour. When 5 per cent of

of the ash of cordite, and obtained a residue among which, under the microscope, crystalline particles could be distinguished. Some of these particles, from their crystalline appearance and double refraction, were silicon carbide; others were probably diamonds. The whole residue was dried and fused at a good red heat in an excess of potassium bifluoride, to which was added, during fusion, 5 per cent of nitre. (Previous experiments had shown me that this mixture readily attacked and dissolved silicon carbide; unfortunately it also attacks diamond to a slight degree.) All the operations of washing and acid treatment were performed in a large platinum crucible by decantation (except the preliminary attack with nitric acid and potassium chlorate, when a hard glass vessel was used); the final result was washed into a

silica was added to cordite, the residue of the closed vessel explosion contained a much larger quantity of these spheres.

shallow watch-glass and the selection made under the microscope. The residue, after thorough washing and then heating in fuming sulphuric acid, was washed, and the largest crystalline particles picked out and mounted.

From the treatment the residual crystals had undergone, chemists will agree with me that diamonds only could stand such an ordeal; on submitting them to skilled crystallographic authorities my opinion is confirmed. Speaking of the largest crystal, one eminent authority calls it "a diamond showing octahedral planes with dark boundaries due to high refracting index." After careful examination, another authority writes of the same crystal diamond, "I think one may safely say that the position and angles of its faces, and of its cleavages, the absence of birefringence, and the high refractive index are all compatible with the properties of the diamond crystallising in the form of an octahedron. Others of the remaining crystals; which show a similar high refractive index, appeared to me to present the same features."

It would have been more conclusive had I been able to get further evidence as to the density and hardness of the crystals ; but from what I have already said I think there is no doubt that in these closed vessel explosions we have another method of producing the diamond artificially.

# CHAPTER X

## THE NATURAL FORMATION OF THE DIAMOND

AN hypothesis is of little value if it only elucidates half a problem. Let us see how far we can follow out the ferric hypothesis to explain the volcanic pipes. In the first place we must remember these so-called volcanic vents are admittedly not filled with the eruptive rocks, scoriaceous fragments, etc., constituting the ordinary contents of volcanic ducts.

Certain artificial diamonds present the appearance of an elongated drop. I have seen diamonds which have exactly the appearance of drops of liquid separated in a pasty condition and crystallised on cooling. Diamonds are sometimes found with little appearance of crystallisation, but with rounded forms similar to those which a liquid might assume if kept in the midst of another liquid with which it would not mix. Other drops of liquid carbon retained for sufficient time above their melting-point would coalesce with adjacent drops, and on slow cooling would separate in the form of large perfect crystals. Two drops, joining after incipient crystallisation, might assume the not uncommon form of interpenetrating twin crystals.

Many circumstances point to the conclusion that the diamond of the chemist and the diamond of the mine are strangely akin as to origin. It is evident that the diamond has not been formed *in situ* in the blue ground. The genesis must have taken place at vast depths under enormous pressure. The explosion of large diamonds on coming to the surface shows extreme tension. More diamonds are found in fragments and splinters than in perfect crystals; and it is noteworthy that although these splinters and fragments must be derived from the breaking up of a large crystal, yet in only one instance have pieces been found which could be fitted together, and these occurred at different levels. Does not this fact point to the conclusion that the blue ground is not their true matrix? Nature does not make fragments of crystals. As the edges of the crystals are still sharp and unabraded, the *locus* of formation cannot have been very distant from the present sites. There were probably many sites of crystallisation differing in place and time, or we should not see such distinctive characters in the gems from different mines, nor indeed in the diamonds from different parts of the same mine.

I start with the reasonable supposition that at a sufficient depth * there were masses of molten iron at great pressure and high temperature, holding carbon in solu-

* A pressure of fifteen tons on the square inch would exist not many miles beneath the surface of the earth.

tion, ready to crystallise out on cooling. Far back in time the cooling from above caused cracks in superjacent strata through which water * found its way. On reaching the incandescent iron the water would be converted into gas, and this gas would rapidly disintegrate and erode the channels through which it passed, grooving a passage more and more vertical in the necessity to find the quickest vent to the surface. But steam in the presence of molten or even red-hot iron, liberates large volumes of hydrogen gas, together with less quantities of hydrocarbons † of all kinds—liquid, gaseous, and solid. Erosion commenced by steam would be continued by the other gases; it would be easy for pipes, large as any found in

* There are abundant signs that a considerable portion of this part of Africa was once under water, and a fresh-water shell has been found in apparently undisturbed blue ground at Kimberley.

† The water sunk in wells close to the Kimberley mine is sometimes impregnated with paraffin, and Sir H. Roscoe extracted a solid hydrocarbon from the "blue ground."

South Africa, to be scored out in this manner.

Sir Andrew Noble has shown that when the screw stopper of his steel cylinders in which gunpowder explodes under pressure is not absolutely perfect, gas escapes with a rush so overpowering and a temperature so high as to score a wide channel in the metal. To illustrate my argument Sir Andrew Noble has been kind enough to try a special experiment. Through a cylinder of granite he drilled a hole 0·2 inch diameter, the size of a small vent. This was made the stopper of an explosion chamber, in which a quantity of cordite was fired, the gases escaping through the granite vent. The pressure was about 1500 atmospheres and the whole time of escape was less than half a second. The erosion produced by the escaping gases and by the heat of friction scored out a channel more than half an inch diameter and melted the granite along the course. If steel and granite are thus vulnerable at comparatively moderate gaseous pressure, it is easy to imagine the destructive upburst of hydrogen and water-gas, grooving for itself a channel in the diabase and quartzite, tearing fragments from resisting rocks, covering the country with debris, and finally, at the subsidence of the great rush, filling the self-made pipe with a water-borne magma in which rocks, minerals, iron oxide, shale, petroleum, and diamonds are violently churned in a veritable witch's cauldron! As the heat abated the water vapour would gradually give place to hot water, which, forced through the magma, would change some of the mineral fragments into the existing forms of to-day.

Each outbreak would form a dome-shaped hill; the eroding agency of water and ice would plane these eminences until all traces of the original pipes were lost.

Actions such as I have described need not have taken place simultaneously. As there must have been many molten masses of iron with variable contents of carbon, different kinds of colouring matter, solidifying with varying degrees of rapidity, and coming in contact with water at intervals throughout long periods of geological time—so must there have been many outbursts and upheavals, giving rise to pipes containing diamonds. And these diamonds, by sparseness of distribution, crystalline character, difference of tint, purity of colour, varying hardness, brittleness, and state of tension, have the story of their origin impressed upon them, engraved by natural forces—a story which future generations of scientific men may be able to interpret with greater precision than is possible to-day.

# CHAPTER XI

## METEORIC DIAMONDS

SENSATIONAL as is the story of the diamond industry in South Africa, quite another aspect fixes the attention of the chemist. The diamonds come out of the mines, but how did they get in? How were they formed? What is their origin?

Gardner Williams, who knows more about diamonds than any man living, is little inclined to indulge in speculation. In his fascinating book he frankly says:

"I have been frequently asked, 'What is your theory of the original crystallisation of the diamond?' and the answer has always been, 'I have none; for after seventeen years of thoughtful study, coupled with practical research, I find that it is easier to "drive a coach and four" through most theories that have been propounded than to suggest one which would be based on any non-assailable data.' All that can be said is that in some unknown manner carbon, which existed deep down in the internal regions of the earth, was changed from its black and uninviting appearance to the most beautiful gem which ever saw the light of day."

Another diamond theory appeals to the imagination. It is said the diamond is a gift from Heaven, conveyed to earth in meteoric showers. The suggestion, I believe, was first broached by A. Meydenbauer,* who says, "The diamond can only be of cosmic origin, having fallen as a meteorite at later periods of the earth's formation. The available localities of the diamond contain the residues of not very compact meteoric masses which may, perhaps, have fallen in prehistoric ages, and which have penetrated more or less deeply, according to the more or less resistant character of the surface where they fell. Their remains are crumbling away on exposure to the air and

* *Chemical News*, vol. lxi, p, 209, 1890.

sun, and the rain has long ago washed away all prominent masses. The enclosed diamonds have remained scattered in the river beds, while the fine light matrix has been swept away."

According to this hypothesis, the so-called volcanic pipes are simply holes bored in the solid earth by the impact of monstrous meteors—the larger masses boring the holes, while the smaller masses, disintegrating in their fall, distributed diamonds broadcast. Bizarre as such a theory appears, I am bound to say there are many circumstances which show that the notion of the heavens raining diamonds is not impossible.

The most striking confirmation of the meteoric theory comes from Arizona. Here, on a broad open plain, over an area about five miles in diameter, have been scattered one or two thousand masses of metallic iron, the fragments varying in weight from half a ton to a fraction of an ounce. There is no doubt these masses formed part of a meteoric shower, although no record exists as to when the fall took place. Curiously enough, near the centre, where most of the meteorites have been found, is a crater with raised edges three-quarters of a mile in diameter and about 600 feet deep, bearing exactly the appearance which would be produced had a mighty mass of iron struck the ground and buried itself deep under the surface. Altogether, ten tons of this iron have been collected, and specimens of the Canyon Diablo meteorite are in most collectors' cabinets.

An ardent mineralogist—the late Dr. Foote—cutting a section of this meteorite, found the tools were injured by something vastly harder than metallic iron. He examined the specimen chemically, and soon after announced to the scientific world that the Canyon Diablo meteorite contained black and transparent diamonds. This startling discovery was afterwards verified by Professors Moissan and Friedel, and

Moissan, working on 183 kilogrammes of the Canyon Diablo meteorite, has recently found smooth black diamonds and transparent diamonds in the form of octahedra with rounded edges, together with green, hexagonal crystals of carbon silicide. The presence of carbon silicide in the meteorite shows that it must at some time have experienced the temperature of the electric furnace. Since this revelation the search for diamonds in meteorites has occupied the attention of chemists all over the world.

Fig. 23 A, C, and D, are reproductions of photographs of true diamonds I myself have extracted from the Canyon Diablo meteorite.

Under atmospheric influences the iron would rapidly oxidise and rust away, colouring the adjacent soil with red oxide of iron. The meteoric diamonds would be unaffected and left on the surface of the soil, to be found haphazard when oxidation had removed the last proof of their celestial origin. That there are still lumps of iron

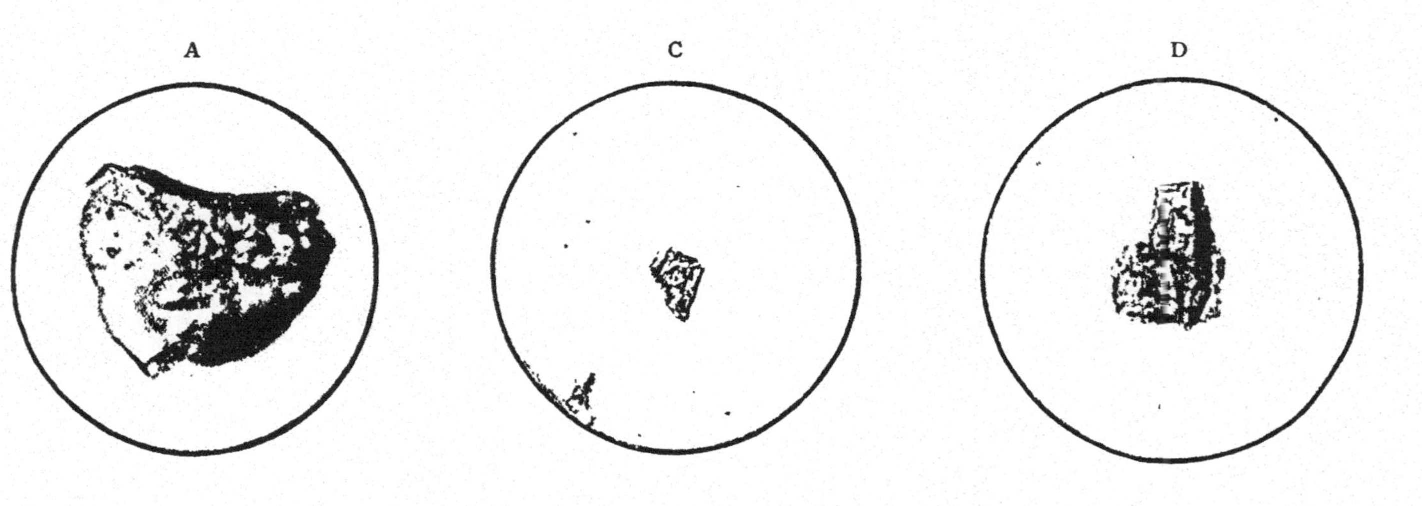


FIG. 23. DIAMONDS FROM CANYON DIABLO METEORITE.

To face p. 138

left at Arizona is merely due to the extreme dryness of the climate and the comparatively short time that the iron has been on our planet. We are here witnesses to the course of an event which may have happened in geologic times anywhere on the earth's surface.

Although in Arizona diamonds have fallen from the skies, confounding our senses, this descent of precious stones is what may be called a freak of nature rather than a normal occurrence. To the modern student of science there is no great difference between the composition of our earth and that of extra-terrestrial masses. The mineral peridot is a constant extra-terrestrial visitor, present in most meteorites. And yet no one doubts that peridot is also a true constituent of rocks formed on this earth. The spectroscope reveals that the elementary composition of the stars and the earth are pretty much the same; and the spectroscope also shows that meteorites have as much of earth as of heaven in their composition. Indeed, not only are the selfsame elements present in meteorites, but they are combined in the same way to form the same minerals as in the crust of the earth.

It is certain from observations I have made, corroborated by experience gained in the laboratory, that iron at a high temperature and under great pressure—conditions existent at great depths below the surface of the earth—acts as the long-sought solvent for carbon, and will allow it to crystallise out in the form of diamond. But it is also certain, from the evidence afforded by the Arizona and other meteorites, that similar conditions have existed among bodies in space, and that on more than one occasion a meteorite freighted with jewels has fallen as a star from the sky.

# INDEX

Able, Sir F., closed vessel experiments, 122
Absorption spectrum of diamond, 101
Aliwal North, 6
Alluvial deposits of diamonds, 9
Amygdaloidal trap, 10
Arizona meteor, 136
Arkansas, diamonds in, 2
Ash of diamond, 82, 89
Augite, 20
Automatic diamond collector, 56

Barytes, 71
— density of, 93
Basalt, 15
Basutos, 12, 39
Bechuanas, 12, 39
Beryl, density of, 93
— refractive index of, 103
Biotite, 20
Blackening of diamonds, 98
Blue ground, 10, 47
— — diamantiferous, 18, 19
Boart, 81
— combustion temperature of, 90
— density of, 93
Boiling-point of carbon, 110
Bonney, Rev. Professor, 67
Boyle on the diamond, 100
Brazil, diamonds in, 4
Breakwater, Cape Town, 36
Breccia, diamantiferous, 19
Brilliant cut diamond, 102
British Association in South Africa, 7
British Guiana, diamonds in, 4
Bronzite, 20, 71
— hydrated, 19
Bultfontein Mine, 14
— — characteristics of diamond from, 64
Bursting of diamonds, 105

Calcite, 20, 97
California, diamonds in, 3
Canada balsam, refractive index of, 103
Canyon Diablo meteorite, 136
Cape Colony, 5
Cape Town, 5
Carat, equivalent in grains, 69
Carbon, boiling and melting point of, 110
— combustion temperature of, 90
— critical point of, 110
— density of, 93
— dissolved in iron, 116
— volatilisation of, 115
Carbonado, 81
— density of, 93

# DIAMONDS

Characteristics of diamonds from the different mines, 64
Chemical properties of diamond, 89
Chromate of lead, refractive index of, 103
Chrome diopside, 71
— iron, 20
— — ore, 71
— — — density of, 93
Chromite, 20
Classification of rough diamonds, 73
Cleavage of diamonds, 78
Coke, density of, 93
Colesberg Copje, 26
Collecting the gems, 55
Coloured diamonds, 62, 82
Combustion of diamond, 89
— temperatures of diamond, boart, graphite, and carbon, 90
"Comet" crushers, 49
Compound system, 36, 37
Concentrating and washing machinery, 49
Convict labourers, 71
Cordite, diamond from explosion of, 123
Corundum, 20
— density of, 93
Cradock, 6
Craters or pipes, 18
Crown glass, refractive index of, 103
Crusher, "Comet," 49
Crystallisation of diamond, 86
Crystals, octahedra, of diamond, 63, 86
Cullinan diamond, 15, 76, 80, 104

Dallas, Captain, 40
DeBeers Consolidated Mines, 7, 33
— — floors at Kenilworth, 47
— — Mine, 14, 24, 34
— — — characteristics of diamonds from, 64
— — strong-room, 74
Delhi diamond, 107
Density of diamond, 57, 93
— of graphite, 83, 93
— of stones accompanying diamond, 70, 71, 93, 95
Depositing floors, 46
Dewar, Sir J., conversion of diamond into graphite, 123
Diabase, olivine, 16
Diallage, 20
Diamond, absorption spectrum of, 101
— and polarised light, 104
— a new formation of, 122
— ash of, 82, 89
— collector, automatic, 56
— combustion of, 89
— — temperature of, 90
— converted into graphite, 100
— density of, 57, 93
— etched by burning, 88
— explosion of, 120
— genesis of the, 115
— in meteors, 134
— in Röntgen rays, 107
— matrix of, 67
— natural formation of, 127
— Office at Kimberley, 73

# INDEX

Diamond, physical and chemical properties of, 89
— pipes or craters, 18
— radio-activity of, 109
— refractive index of, 103
— Trade Act, 36
— triangular markings on, 87
— tribo-luminescence of, 100
Diamonds, coloured or fancy, 62, 82
— Maskelyne on, 1
— noteworthy, 76
— phosphorescence of, 96
— produced, weight, value of, 35
— yield of, from De Beers, 60
Drift, diamonds from the, 12
Duke of Tuscany diamond, 80
Dutch boart, or zircon, 59
Dutoitspan Mine, 14, 23
— — characteristics of diamonds from, 64

Eclogite, 20
— containing diamonds, 67
Electrons, bombardment by, 98
Emerald, refractive index of, 103
Empress Eugenie diamond, 80
Enstatite, 20
Explosion of diamonds, 120
Excelsior diamond, 80

Fancy stones, 62
Fingoes, 39
Flint glass, refractive index of, 103
"Floating Reef," 21
Floors, depositing, 46
Fluor-spar, refractive index of, 103
Formation, new, of diamond, 122
Fort Beaufort, 6
Franklinite, 97
Frank Smith Mine, 15
— — — characteristics of diamonds from, 66
Fraserburg, 6

Garnet, 20, 70
— density of, 93
Genesis of the diamond, 115
"Golden fancies," 65
Granite, 18
— density of, 93
Graphite, 81, 83
— combustion temperature of, 90
— conversion of diamond into, 100
— density of, 93
— diamonds coated with, 99
Graphitic oxide, 83, 93
Grease, collecting diamonds by aid of, 57

Hard blue ground, 47
Hardness of diamond, 90
Haulage system, 46
Hexakisoctahedron crystal, 86
Hope blue diamond, the, 80
Hornblende, 71
— density of, 93

Iceland spar, refractive index of, 103
Ice, refractive index of, 103
I.D.B. laws (Illicit Diamond Buying), 36
Ilmenite, 20
India, diamonds in, 4
Inverel diamonds, 91
Internal strain in diamonds, 104
Iron a solvent for carbon, 116
— ore, density of, 93
— pyrites, 20

Jagersfontein diamond, 79
— Mine, 14
— — characteristics of diamonds from, 68
Jeffreysite, 20

Kafirs, 42
Kamfersdam Mine, 15
— — characteristics of diamonds from, 66
Kenilworth depositing floors, 47
Kimberley, 6
— blue ground, 10
— mines, 14, 23, 34
— Mine in old days, 25
— — at the present day, 34
— — characteristics of diamonds from, 63
— shales, 15
— West Mine, 15
— — — characteristics of diamonds from, 66
Kirsten's automatic diamond collector, 57
Klipdam, 8, 23
Koffyfontein Mine, 14
Koh-i-noor diamond, 80
— hardness of, 91
Kyanite, 20, 71

Lamp, ultra-violet, 97
Leicester Mine, 15, 23
— — characteristics of diamonds from, 67
Loterie d'Angleterre diamond, 80
Lustre of rough diamonds, 56

Machinery for washing and concentrating, 49
Macles, 86
Magnetite, 20, 71
— density of, 93
Maskelyne on diamonds, 1
Matabele, 12, 39
Matrix of diamond, 67
Melaphyre, 10, 16
Melting-point of carbon, 110
Meteor, Canyon Diablo, 136
Meteoric diamonds, 134
Meydenbauer on meteoric diamonds, 135
Mica, 20, 71
— density of, 93
Moissan's experiments on the genesis of diamond, 115
Mud volcano, 24

Nassak diamond, 80
Natal, coal in, 6
Natural formation of diamond, 127
Newlands Mine, 15
— — characteristics of diamonds from, 67

# INDEX

New Rush diggings, 26
Nizam of Hyderabad diamond, 80
Noble, Sir A., experiments, 122, 131
Noteworthy diamonds, 76

Octahedral crystals of diamond, 63, 86
Olivine, 20
— diabase, 16
Orange River Colony, coal in, 6
— — — diamonds in, 14
Orloff diamond, 80

Pasha of Egypt diamond, 80
Paterson, Mr., description of Kimberley in old days, 25
Peridot, 20, 139
Peridotite, 3
Perofskite, 20
Phosphorescence of diamonds, 96
Phosphorus, refractive index of, 103
Physical properties of diamond, 89
Picking tables, 51
Pipes or craters, 18
Pitt diamond, 80
Polarised light and diamond, 104
Pole Star diamond, 80
Pondos, 39, 42
Premier Mine, 15, 76
Prodigious diamonds, 76
Pseudobrookite, 20
Pulsator, 52
Pyrope, 70

Quartzite, 16, 20
— density of, 93
— refractive index of, 103

Radio-activity of diamond, 109
Radium, action on diamond, 108
"Reef," 21
Refractive indices, 103
Refractivity of diamond, 102
Regent diamond, 80
Reunert, Mr., description of Kimberley Mine, 30
Rhodes, Cecil John, 34
River washings, 7
Rock shafts, 43
Röntgen rays, diamond in, 107
Ruby, refractive index of, 103
Rutile, 20

Sahlite, 20
Sancy diamond, 80
Savings of the native workmen, 41
Scalenohedron diamond crystal, 86
Serpentine, 19
Shafts, rock, 43
Shah diamond, 80
Shales, Kimberley, 15
Shangains, 39
Shells in blue ground, 21
Shot boart, 81
Silver and thallium, nitrate of, 94
Smaragdite, 20
Soft blue ground, 47

Sorting the diamantiferous gravel, 55
Specific gravity, *see* Density
Spectrum, absorption of diamond, 101
Sphalerite, 100
Spinthariscope, 108
Sprat's *History of the Royal Society*, 1
Sprouting graphite, 84
Star of the South diamond, 80
Stones other than diamonds, 70, 71, 93, 95
Strain, internal, in diamonds, 104
Sulphur, refractive index of, 103
Swazis, 39

Ultra-violet lamp to show phosphorescence, 97
Underground workings, 43
United States, diamonds in, 2

Vaalite, 20
Vaal River, 8, 16
Valuators, 73
Value of diamonds per carat, 12, 69
Value of diamonds, progressive increase in, 69
Vermiculite, 20
Volatilisation of carbon, 115
Volcanic necks, 18
Volcano, mud, 24

Wages, scale of, 35
Washing and concentrating machinery, 49
Wesselton Mine, 14, 15, 23, 35
— — characteristics of diamonds from, 65
Willemite, 97
Wollastonite, 20
Workings, underground, 43

Yellow ground, diamantiferous, 19
Yield of diamonds, annual, 60
— — — total, 35
— falls off with depth, 68
— per load of blue ground, 62

Zimbabwe ruins, 40
Zircon, 20, 59, 71
— density of, 93
Zulus, 12, 39, 40

W. BRENDON AND SON, LTD., PRINTERS, PLYMOUTH

CPSIA information can be obtained
at www.ICGtesting.com
Printed in the USA
BVOW08s2009070517
483404BV00001B/81/P